# Britain's Nuclear Waste: Safety and Siting

Stan Openshaw
Steve Carver
John Fernie

**Belhaven Press**
A division of Pinter Publishers
London and New York 1294

First published in Great Britain in 1989 by
Belhaven Press (a division of Pinter Publishers).
25 Floral Street, London WC2E 9DS

Reprinted 1989

**British Library Cataloguing in Publication Data**
A CIP catalogue record for this book is available from the
British Library

ISBN 1 85293 001 2 (Hardback)
ISBN 1 85293 005 5 (Paperback)

**Library of Congress Cataloging-in-Publication Data**

Openshaw, Stan.
    Britain's nuclear waste : siting and safety / Stan Openshaw, Steve
Carver, John Fernie.
        p.    cm.
    Bibliography: p.
    Includes index.
    ISBN 1–85293–001–2 (Hardback)
    ISBN 1–85293–005–5 (Paperback)
    1. Radioactive waste disposal—Great Britain.    Radioactive
wastes—Great Britain—Location—Data processing.    I. Carver,
Steve.    II. Fernie, John.    III. Title
TS898.13.G7064    1989
363.72'87'0941—dc20                                    80–14859
                                                        CIP

Photoset in North Wales by Derek Doyle & Associates, Mold, Clwyd.
Printed and bound in Great Britain by Billing and Sons Ltd, Worcester.

# Contents

# Preface

Two of the authors (Stan Openshaw and John Fernie) submitted evidence to the House of Commons Environment Committee during Session 1985–6 whilst it was investigating the radioactive waste problem in Britain. We wanted to draw the Committee's attention to the possible use of computer techniques for identifying and evaluating potential radioactive waste disposal sites (Openshaw and Fernie, 1986). We wrote, 'Decisions in this field will be unpopular, and public opinion concerning the waste management strategy will only be carried if NIREX can show, in public, that it has undertaken a detailed and rigorous evaluation of all possible locations' (p.645). We believed then, as now, that computer geographic methods could be most useful for this purpose and set about preparing this book as an attempt to bring some basic geographic technology to bear on this most difficult problem.

It is useful to start this task by taking a broadly based retrospective and prospective look at the stage the radioactive waste 'story' has reached and then justify why this subject is of such interest. Make no mistake. The radioactive waste disposal problem is not a once and for all time issue, rather it is the beginning of a probably long trail of related matters stretching (according to some views of the future) at least into the 22nd and 23rd centuries; it is not, by its nature, a one-off issue that once solved never recurs again. The radioactive waste problem as such will never be solved because it is likely that there will always be one. The problem of radwaste disposal exists now and may well exist at all subsequent moments in time, or at least until that time when technology has moved on, out of the fission era into fusion, and whatever follows that; the radioactive waste disposal issue is likely to be with us for as long as human civilisation continues.

Does this seem to be a horrifying thought? Really there is relatively little to worry about. The principal problem is probably the manner by which the nuclear era has been introduced. Atomic explosions, scares about invisible radiation damage to health, the occurrence of reactor accidents, an international scene conditioned to accept mutually assured destruction as a way of ensuring peace on an everyday basis, do little to instil public confidence in nuclear technology. This problem is exacerbated when the job of reassuring the public is vested in a nuclear science elite who are totally unable to explain their technology to the man in the street and who operate through political lobbies and via the 'faith principle' (if you do not understand then you have no choice but to trust in us) to survive, and a nuclear 'science' that can by its very nature never provide cast iron assurances and guarantees of safety and which uses probability to explain how small are the relative risks of atomic death compared with n-thousand other ways of dying in the game of life. Given this environment, super-charged with excess emotion, massive ignorance, institutional arrogance, mis-information and political manipulation, what chance is there that the subject of radioactive waste dumps will receive the serious, low key, emotion free, sensible and rational treatment it requires? Many of the hard-line anti-nuclear groups are very wrong but so too are the equally extreme pro-nukes. The latter, it might be argued, are a more dangerous bunch because of their access to power. However, it is important to understand that there is no conspiracy, merely a strongly motivated group of nuclear interested people in key positions who honestly and sincerely believe that an all nuclear future is the only future, and that it is essential to start this process as soon as possible. Once started, you do not (indeed maybe cannot) turn back. Who are we to say that this invisible body of nuclear worthies, eminent scientists, famous politicians and many others, is wrong! We cannot be so bold nor would we necessarily disagree with their assessment. At the same time, it is often quite difficult to accept that what has been happening in the recent past is what should be happening today. Maybe the invisibles are too invisible and have failed to carry the debate to the people; perhaps they are being too unwilling to 'risk' a 'bad response'. If people 'misunderstand' their vision and the need, then maybe they still might be able to terminate the process by democratic action. However, if this 'misunderstanding' occurs too late for termination to be an option, then it no longer matters what people believe – they would no longer have any choice.

Nuclear waste is also part of the wider issue which involves the level of nuclear presence that will be needed in the future in order to provide the energy that modern civilization is thought to require. The current debate for renewable energy supplies is also, incidentally, a powerful argument for a closed nuclear fuel cycle based on the Fast Breeder Reactor. Quite simply, if you want to avoid the problems of dealing with radioactive waste then for many countries it is already too late. You cannot expect to have nuclear weapons, nuclear power, nuclear medicine and other aspects of the nuclear industry without also having a nuclear waste policy, a nuclear waste strategy and at least a few nuclear waste dumps. Few would question that there are benefits from the nuclear age, or that the contribution of nuclear electricity, currently small, will become far more critical in the near future. Indeed, depending on how important they consider certain nuclear-related products to be, most people might well reluctantly agree that the benefits may even be considered substantial and increasingly indispensable to modern life. Growing fears about 'greenhouse' effects, about environmental pollution and about manmade diseases resulting from combustion of fuel are also arguments for a greater reliance on nuclear power.

Enter a typical late 1980s nuclear exhibition. Thunderous martial music, images of power, and a multi-media feeling of awe, and also relief, that the nuclear age is dawning and all the problems resulting from Victorian industrialization will soon be solved. The story continues with a touch of reality. You cannot expect to enjoy all the benefits of the nuclear era without also having to cope with some minor problems. Radioactive waste contains most of the nasties that would otherwise have been released into the environment. Instead, it is 'contained' and will be prevented from entering the environment until it cannot do any harm. More music, Maestro, preferably of a very strident, stirring and confidence creating kind! The Sellafield Exhibition Centre is a classic example that should be visited and studied. It is an intense store of information mixed with propaganda, of audio-visual stimuli, with an unfettered, blinkered and strongly nuclear view of the world. Well done British Nuclear Fuels!

The reality is perhaps different and certainly more mundane. There is no denying the long term importance of nuclear power or that it is currently the only practicable, large scale, long lasting, and virtually pollution free power source known to mankind. At a time when the combustion of fossil fuels are creating the prospect of global climatic change, it certainly cannot be ignored as a possible civilization saving technology in the longer term; although there is an argument that maybe it is already too late to avoid doomsday. Nuclear power cannot be developed quickly enough and its potential greenhouse effect reducing impact will become noticeable when Britain and most other countries of the world all have large scale nuclear power programmes in place. That will take about 100 years, but by then it may be too late. Nevertheless, there is certainly a strong case for the government and key international agencies to start this process rolling by explaining the need and the concern to ordinary people. It cannot be done by the backdoor. It is not a 'softly-softly' job and there needs to be a large degree of international scientific agreement about the seriousness of the situation before this strategy will work. All this needs to be done very soon, otherwise there is a real risk, in Britain at least, that an anti-nuclear backlash will become so vote winning that party politics will result in nuclear power being mothballed; it cannot now be easily scrapped. We are neither anti-nuclear nor are we in the pay of the nuclear industry, but as geographers we sincerely believe that nuclear power is becoming essential to the future of Britain and the world. We do not know how it can now be avoided. Yet it is also clear that in general both the government and the nuclear industry have constantly failed to grasp the key elements of their own arguments. It is true that the reality of the situation has appeared late, and thus smacks of retrospective or *post hoc* justification, but that is only because of the earlier ignorance and blinkered arrogance. Maybe they are less convinced than we are! The basic theme here is that you really must trust the people, take the arguments direct to them and then win the ensuing debate – a strategy that might be termed the Walter Marshall approach! You must not seek to build what amounts to a future critical industry on foundations of sand, when a rock base could have been used. The 'difficulties' in 'handling' the radwaste disposal problem are a classic reflection of the wider difficulties in articulating the nuclear option at a time when it is simultaneously publicly unpopular and future essential.

Why not be honest and frank? Nuclear waste exists not because it was created by God or because it was all produced by immediately and universally beneficial research, not because it was unavoidable, but because it was actually, astonishing as it may seem, quite deliberate. There is also a difference between the fact of its existence and the *amount* that exists. Small amounts exist because we have nuclear weapons to preserve the peace. Since such weapons are apparently regarded as absolutely essential to our country, presumably these wastes should be cherished as successful (so far) eliminators of world war. As such perhaps they could be incorporated into peace monuments or divided into pea-sized lead encased presents with one given to every family: a suggestion originally made as a solution to the high level waste problem! After a few hundred years, people dig it out and stare at it (although only for a few minutes) in awe and wonder, and say 'thanks' for all those years of peace. A much larger amount, maybe equivalent to an orange per family in size, has been deliberately created by a desire to reprocess nuclear fuel. It is not necessary to reprocess nuclear fuel; for example, there is no reprocessing of civilian nuclear fuel in the United States, but for various reasons the decision was made to do so. At the time, *circa* mid-1950s, it made sense and was probably the only course of action that was considered possible in the United Kingdom, partly because of the nature of British nuclear power stations and partly because of a thirst for plutonium. The corollary of reprocessing nuclear fuel is that a much larger amount of radioactive waste is created. If you did not reprocess, then there would still be waste to dispose of, albeit in much smaller amounts. The disposal problem would still occur but probably much later as it could be hidden for a much longer period of time. Even without a nuclear power programme and nuclear weapons, there would still be some nuclear wastes but in miniscule amounts. Of course, this waste-generating activity by reprocessing is not completely ridiculous. Reprocessing spent fuel is necessary to close the nuclear fuel cycle and is widely regarded as essential in order to conserve uranium supplies and provide virtually unlimited power supplies. The view might be taken that the decision to do this in Britain was made possibly 100 years before it would seem essential. However, in the mid-1950s this need was real and also necessary to prove it could be done. Of course, once you start down a reprocessing route it is difficult to stop and, moreover, many people see no reason to. Once you have to invest in the necessary infrastructure it might as well be used.

So whether we now like it or not we have a heap of nuclear waste to dispose of in either a repository (a retrieval store) or a depository (non-retrieval store);

we prefer to call the latter a radwaste dump, the term tip conjures up the wrong impression! The wastes we now have and much larger amounts that we are now committed to having, are a result of decisions already made. It is now too late to decide not to have as much, or any! The waste has to be dumped or stored or disposed of somewhere, somehow, preferably soon, probably within the next fifty to one hundred years or so. The amount of additional waste that might appear in the future is dependent on decisions not yet made and thus the true magnitude of the radwaste problem in 2030AD is still an unknown and uncertain quantity. What is certain, is that by 2130AD there will be one, maybe three (or more), national radwaste dumps.

Despite the 'we have got to have at least one dump soon' argument, it is difficult to imagine a worse problem than the disposal of radioactive wastes. The rubbish exists so you have got to put or dump it somewhere, and the fact that it is radioactive and that most of it is likely to remain so for long (some for extremely long) periods of time is seemingly quite exceptional. However, it is not really that special or unusual. Mankind has been dealing with hazardous substances for about 3000 years, usually badly and in an immensely careless manner; you can still detect the effects of the Roman lead miners and no doubt people are still ingesting some of the lead the Romans mined. So why all the fuss about radioactive waste, especially when it is clear that considerably more care and thought than usual is going to be taken to dispose of the most hazardous bits?

Perhaps, the real problem is that people feel trapped by the inevitability of it all. Whether we like it or not, Britain has today a sizeable nuclear waste disposal problem and, regardless of people's attitudes for or against radioactivity or nuclear power, something will have to be done about it fairly soon. We have been led into a situation where suddenly there is no choice. A heap of waste exists; there is a seemingly unalterable committtment to much more in the near future – a decision is needed soon. It is possible to call those responsible criminally insane, but such a perspective is not helpful even if it does make some people feel better. The problem that has now been 'found' to exist could have been predicted with almost complete accuracy in the late 1950s, if anyone had wanted to. At that time, there was no concept of public acceptability as an important factor in nuclear decision making. Due to secrecy, such matters were neither in the public nor political arenas. Maybe also, most of the waste was simply dumped at sea. The task now is how to 'rescue' the nuclear industry from the problem of their own predecessors' making while recognizing that any solution is important to all of us. We are trying to meet the challenge by seeking to open up the debate and to provide an independent geographical view of the disposal of radioactive waste, outlining how it might best be handled. We try and do this by setting up 'Aunt Sally's'. By being provocative to both sides it may be possible to 'draw the venom' out of the debate and thereby provide a basis for an acceptable pragmatic solution to a very important problem.

The book is also an attempt to inform an interested reader about what has been happening and to offer a commentary on the actions of both government departments and the nuclear industry. It is written in a language that ordinary members of the public, elected representatives, government ministers, and anyone else who may be interested should be able to understand. Indeed, an advantage enjoyed by geography is that it can often be used to communicate a complex problem in a way that ordinary people are better able to appreciate. This is a book for both the man in the street and the nuclear expert. The purpose is not to take sides but to present an independent and informed view of the current state of play in a game that is important to us all.

The key issue was and still is a purely geographical matter which transcends the scientific debate. WHERE DO YOU PUT YOUR FIRST NATIONAL RADWASTE DUMP? In the ongoing debate, information about the full set of possible locations is obviously of considerable importance as a means of testing whether the 'best' site(s) have been selected and that the national interest is best served. Some may argue that this book is now too late. It 'appears' (in early 1989 when these words were written) that the 'key' siting decision has been made and that therefore there is nothing left to do, or argue about. This particular 'key' decision has already been made on at least two previous occasions and then reversed, so it is never too late to try and influence the outcome and there is still plenty of opportunity to do so. A number of major stages in the decision making sequence have still to be reached. All we have at present is a statement of intent and a decision in principle as to how to proceed. Whether the preferred solution is considered acceptable has yet to be tested. There is a prospect of a lengthy public inquiry and probably at least one general election before any major decision will be made, even then it could conceivably be reversed at a still later date. The eventual existence of a radwaste dump is not in doubt, but the date when one becomes operational is much more uncertain. No government is going to stand or fall on its handling of radioactive waste and every member of the major parties knows it. Only when one (or more) radwaste facilities have been in full operation for a year or so would it be too late to write this book. It is imagined that this critical stage will not be reached before the early years of the next century. Accordingly, there is plenty of time to ponder, discuss, and debate this most difficult, unattractive, and yet also most important of subjects.

Stan Openshaw
Steve Carver
John Fernie

St Valentine's Day 1989

# Acknowledgements

The authors wish to acknowledge the help and assistance given to them by the following people and institutions whilst preparing this book: Martin Charlton of CURDS, Newcastle University provided valuable computing advice. Dr Terry Ratcliffe and Ron Pringle of NUMAC at Newcastle put up with all the computing. The ESRC assisted in the form of a PhD studentship for Steve Carver, and the North East Regional Research Laboratory provided the GIS infrastructure (aided by the University's Equipment Committee). We are also grateful to the Ordnance Survey for providing access to their 1:625,000 digital map data. The BGS and Countryside Commission provided paper maps, and the ESRC data archive supplied the census data. Finally the population counts were made available by Professor David Rhind of Birkbeck College, London. Thank you.

# 1

# What's all this about radioactive waste?

## The problem?

Let us start with an appraisal of the current situation. It is an unavoidable fact that there now has to be a radwaste facility somewhere in Britain fairly soon, the only real question is where to put it. Answering this question is not as it may seem at first sight, a purely technical and objective scientific process, but one that involves extremely complex and difficult value judgements, combining technical and broader socio-economic and political concerns. Understanding the incredible difficulty and also the immense importance and long term significance of this task involves delving into the history, the policy background and the political context within which the radwaste disposal story has evolved. In this chapter we try and set the scene for those that follow.

*Table 1.1* Estimates of radioactive waste holdings in 1987 (m³)

| Waste Type | raw | conditioned |
| --- | --- | --- |
| Low Level Wastes (LLW) | na | 2,340 |
| Intermediate Level Wastes (ILW) | 43,600 | 59,400 |
| High Level Wastes (HLW) | 1,430 | 517 |

Source: RWMAC, (1988b) p51
Note: LLW are always conditioned

Table 1.1 shows estimates of the current stocks of radioactive wastes. The various categories are described in detail in chapter 2; it is sufficient to say here that the wastes of interest are the low level wastes (LLW) and the intermediate level wastes (ILW). The third category of high level wastes (HLW) is currently being handled by storage for the next 50 years and it is governmental policy to concentrate on the LLW and the ILW first. It is also government policy that these wastes are to be disposed of under strict supervision to high safety standards and that the period of storage should be the minimum compatible with safe disposal. The apparently small size of the LLW stocks in 1987 reflects the use of the Drigg dump (see Chapter 5). The large stocks of ILW partly result from the suspension of sea dumping in 1983 and the consequential storage at Harwell awaiting a disposal route.

*Table 1.2* Sources of radioactive waste holdings in 1987 (%)

| Source | Waste category | | |
| --- | --- | --- | --- |
| | LLW | ILW | HLW |
| Sellafield | 23 | 75 | 86 |
| Other nuclear industry | 40 | 14 | 0 |
| Research, medical, industrial | 36 | 11 | 14 |

Source: RWMAC (1988b) p.53

The obvious next question is where did the wastes come from. The vast majority of current stocks is a result of nuclear fuel reprocessing. Table 1.2 gives a breakdown by major sources. However, this table is misleading because it artificially separates 'other nuclear industry' from 'research, medical, industrial'. This would appear to imply that a large proportion of wastes is being generated outside of the nuclear industry, thereby presumably increasing the scope of the problem and hence broadening the support for a solution. In reality, virtually all the ILW and HLW come from the nuclear industry either at Sellafield or from its reactors or its research organisations. The medical and non-nuclear industrial sources are, by comparison, very minor generators of wastes. The LLW picture is similarly distorted by a massive under-representation of Sellafield's contribution. On this occasion, it is due to the definition of LLW as being waste that is transported away from the site of origin. At Sellafield most of the wastes go to Drigg and would be excluded. Surely this statistical distortion is unnecessary and serves only to confuse. The real situation is that Sellafield generates about three-quarters of all the non-decommissioning wastes. It is estimated that by 2030AD, 76 per cent of the LLW, 62 per cent of the ILW and 77 per cent of the HLW will have been generated at Sellafield (Environment Committee, 1986; p.xxvi). Most of the current wastes for which disposal is now required result from reprocessing activities. Reprocessing is discussed further in Chapter 2. It is a significant waste generator because it increases the amount of waste by a factor of about 200, compared with the input spent fuel rods. It is estimated that 4 cubic metres of spent fuel when it is reprocessed will produce: 2.5 cubic metres of HLW, 40 of ILW and 600 of LLW.

It is argued that the radioactive waste (radwaste) disposal problem has two components to it: (1) what to do with the existing inventory of radioactive waste which either already exists or is going to fairly soon because of current committments relating to: reprocessing, nuclear weapons, nuclear power stations, nuclear submarines, nuclear medicine (Table 2.1 gives stocks to 2030AD); and (2) that which may be created in the future as a result of decisions not yet made and commitments not yet accepted involving: decommissioning, fuel reprocessing and the size of the nuclear power programme. It would appear that the existing problem when viewed in terms of current arisings and future committments, is already sufficiently large to require more than one national radwaste disposal facility to cover the various types and sources of LLW and ILW wastes. By their nature these two key questions are largely political in that their answer should reflect explicit policy considerations. In reality, it seems that the decisions have been subcontracted to the nuclear industry; presumably, so that the politicians have someone to blame should it be necessary to do so. The questions arise as to who determines the size of the nuclear industry, who decides on the level of reprocessing activity and who should be planning for the radwaste problem. In all cases the answer should be the government, but in all instances the role of government has been restricted to matters of general policy and the detailed decisions left to the nuclear industry. There are arguments for and against this situation, but the public may well be surprised to discover the lack of government control on the key parameters of such a critical industry. For instance, reprocessing British spent fuel is one thing, but having a reprocessing plant three times larger than the British internal market can handle is something else. The British public's radiation dose commitment will, because of this zeal, be much larger, irrespective of whether the foreign fuel wastes are dumped here or abroad. Why should Britain become a nuclear dustbin merely to increase the profits of the nuclear business? Why shouldn't Britain mainly concentrate on handling its own or EEC-generated wastes? The long lag between decisions and waste arisings serves to confuse further nearly all the issues.

*Table 1.3* Reactor decommissioning 1990–2060

| Time Period | Megawatts | Number of reactors | Cumulative number |
|---|---|---|---|
| – 1990 | 332 | 6 | 6 |
| 1991 – 2000 | 3385 | 24 | 30 |
| 2001 – 2010 | 4636 | 8 | 38 |
| 2011 – 2020 | 2640 | 4 | 42 |
| 2021 – 2030 | 2640 | 4 | 46 |
| 2031 – 2040 | – | – | 46 |
| 2041 – 2050 | 2400 | 2 | 48 |
| 2051 – 2060 | 7200 | 6? | 54? |

Notes: MAGNOX and AGRs assumed to have 32-year life; PWRs 40 years
Assumes PWR programme extended to 8 reactors by 2000 with one new greenfield site

The nuclear industry probably hopes that by the time the full extent of the current radwaste committments are widely understood, by both public and politicians, it will be too late to stop the process. The real 'hidden' component is the decommissioning wastes that will not start contributing large amounts until after 2000AD. Table 1.3 gives estimates of the numbers of nuclear reactors likely to be decommissioned by 2060AD; it excludes decommissioning reprocessing plant. While the ultimate size of this future problem can be manipulated, for example, by the institution of more efficient waste management practices and by altering the nature of the decommissioning procedures used, the fact that there will be a 'new' problem due to the need to handle these decommissioning wastes is now both undeniable and also unavoidable. It is estimated that even by 2030AD the total LLW stocks will be increased by 22 per cent and ILW by 34 per cent due to the early stage decommissioning of MAGNOX and AGR (Advanced Gas-cooled Reactor) plant.

The best way to avoid dealing with old reactor hulks is to minimize their impact by keeping old reactors running for as long as possible and then to redevelop the site with more nuclear facilities. If you wish to abandon nuclear power and leave the reactors standing as atomic monuments such will doubtlessly not be viewed as very satisfactory by those for whom the reactor building is the very apotheos of fear and hatred, the same people who want to cancel the nuclear option. Abandoning nuclear power and clearing the sites quickly creates a massive and immediate radwaste problem of considerably greater dimensions than a more leisurely decommissioning process taking place over a century or so would have done. We wonder whether those wanting to abandon nuclear power as an energy option fully realize the long-term nature of the task. The Faustian nature of the nuclear deal is seldom recognized: it is far easier to embrace nuclear power than it is to remove it, if indeed this can ever be done.

On the other hand, it is increasingly clear that nuclear power is probably here to stay. It is the only known energy source able to meet the world's long-term energy needs, and with increasing concern about the climatic impacts of fossil fuel combustion, in particular the greenhouse effect and acid rain, the attractions of a high-nuclear future may well grow. However, the logic of a high nuclear contribution leading gradually and inexorably, to an all nuclear future also requires that fast breeder reactors replace the current thermal reactors (viz the PWRs (Pressurized Water Reactor), AGRs etc). Fast Breeder Reactors (FBR) bring with them the prospect of a plutonium-based world and are only feasible if the nuclear fuel cycle can be closed. This requires that large quantities of irradiated fuel can be reprocessed with the consequential need for a large radioactive waste disposal capability. All this, even on optimistic forecasts, is still about a hundred years away. You cannot switch to FBRs overnight. The initial plutonium fuel loads may well take 20 to 30 years of AGR or PWR spent fuel reprocessing to buildup. The current interest in the reprocessing of 'spent fuel' from thermal reactors is therefore almost certainly not for nuclear weapons, but to establish the plutonium needed for a small FBR programme. Once it starts, provided the nuclear fuel cycle is closed, the numbers of FBRs can be increased fairly quickly. The key to all these future possibilities is the establishment of operational and reasonably acceptable radioactive waste dump(s) somewhere (indeed anywhere) as soon as possible. The barrier is mainly psychological, for until such a facility exists, reactors are being built and fuel reprocessed without any available disposal route for most of the more radioactive wastes that are being created. Sooner or later there will come a time when shortage of temporary storage space and, more seriously, political concern about the lack of a permanent solution, may eventually slow and then stop new nuclear developments. At present, many developments are proceeding on the assumption that these facilities will be in place fairly soon, certainly by the time they are needed. The key time scale is almost certainly occasioned by the 10 – 20 year working life of the thermal oxide reprocessing plant (THORP) at Sellafield which is due to start in the early 1990s, the imminent need to dispose of old bits of nuclear submarine reactors, and the start of decommissioning of the early civilian MAGNOX power stations. Clearly, a solution of some kind is needed soon.

## Psychological aspects for both pro- and anti-nukes

The most critical aspect of the current search for a radioactive waste dump site is that once one is found, the last remaining non-technical barrier to unlimited nuclear expansionism will have gone. It is not that there are now commitments to a massive unlimited nuclear programme, the nuclear planning process operates in a much more piecemeal and disjointed incremental manner. Moreover, in the semantics of the nuclear debate there is no commitment to anything not already contracted for. Rather, the removal of this final barrier will be of considerable symbolic significance. A sign that when necessary, and when the need is there, there are no longer any extra-technical obstacles to be overcome.

It follows, therefore, that in many ways the first dump to be built will be the most difficult of them all. The first pro-nuclear country to succeed in this task will also become a show-case that many others may wish or have to follow. The current lack of examplar elsewhere in the world merely increases the global significance of the current British attempt. It is noted that the recently commissioned Swedish facility is for storage of spent fuel rather than reprocessing wastes. The remaining problems are probably no longer of a technical nature, but are political and mainly a matter of gaining some reasonable level of public acceptability (even if it is just local). 'Stop Radwaste Dump Number 1' and maybe you also simultaneously cripple (but not kill) the nuclear industry of the future (and probably also fundamentally damage future standards of living in Britain). On the other hand, failure here is not unusual and the nuclear industry can afford to be patient for another 50 – 100 years, even if it would be nice to start the ball rolling at a much earlier date.

If a strident anti-nuclear policy is adopted, with the priority closure of nuclear power stations, then a waste disposal problem will, if left to slowly unfold over a long period of time, become concentrated and will have to be dealt with much sooner rather than later; it may also be of considerably greater dimensions. Likewise, if you abandon nuclear weapons and have no reactors able to burn-off the weapons grade plutonium, then again you may create a big problem of short-term disposal which is complicated by proliferation issues; it would be important to deny access to the weapons-grade plutonium. So to some large degree opposition to the concept of radioactive waste disposal is doomed to fail; in the national interest there must be some facilities fairly soon.

In trying to develop an acceptable disposal facility a major difficulty is the scientifically irrational, but equally quite understandable 'fears' that nuclear issues provoke in otherwise quite ordinary and normal people. Quick (1988) puts it rather neatly:

When one considers that low level waste consists of such materials as workmen's gloves and overalls, broken equipment, soil, etc, one begins to wonder what all the fuss is really about ... there could even be a situation in which the soil being excavated to construct a site could have been more radioactive than the material eventually put into the repository. [p.21].

On the other hand, not all waste is so innocuous or so possessed of such small levels of activity. The spectrum of what is considered to be low level waste is fairly wide (see Chapter 2).

Additionally, it is really quite reasonable for people to object to having a waste dump near to their homes and workplaces. Most people object to ordinary waste tips let alone ones handling radioactive substances. A key variable here is some notion of what 'nearby' means. Does nearby mean 100 metres, or the same local authority, or the same drainage basin (or acquifer feed area), or the same part of the country. The fear of being 'nearby' is not a matter of scientific risk or fact, but of some psychological perception of being too close to some unknown and terrible danger (Macgill, 1987). The problem here is that it is these perceived rather than objective scientific risks that tend to influence people's behaviour and decision making processes.

The fears are understandable albeit based on faulty logic. Is education the answer or do you respond to these genuine fears in a sympathetic manner and seek locations where there are as few people as possible 'nearby', with a definition that is not purely scientific but one based on sensitivity to public fears and concerns? Can you 'buy' public acceptance by seeking the remotest of sites and then live with the consequences? The answer is probably 'yes'. But of course it is not neccessary on scientific grounds nor mandatory on political ones, even if it appears no more than commonsense at the level of the man in the street. However, many people working in this area seem to operate using only science-based rationale rather than ordinary commonsense and herein lies the cause of some of the problems.

## Nuclear science and nuclear planning

In Britain the nuclear planners cannot really be accused of going 'soft' to gain public acceptability. Previously, national self-interest (mainly in relation to defence) has ensured a minimal level of accountability to public fears and concerns. Indeed, until the mid-1970s, a combination of secrecy and apathy ensured an easy time. Three Mile Island, hypothetical cancer clusters that appear to be caused by mythical viruses that apparently only live near to nuclear facilities and Chernobyl have changed all that, by making the subject into a media extravaganza. Traditionally, the nuclear industry's pure or hard science view has nearly always won through. The process is more or less as follows. In the early days, there was no debate and the scientists had it all their way apart from the usual random political interference. Since then, an era of increasingly contested public inquiries has become typical. Yet even in this context the problem can often be solved in the following manner. First, it will be claimed that site 'X' has to be used because: (a) it is safe, (b) it is acceptable from an engineering point of view, (c) there are no other sites as good in the region of interest, (d) it is deemed acceptable by the various supposedly nuclear industry independent and public safety watchdogs, and (e) the balance of judgements favours this location in the national interest. A Public Inquiry will be held so that the protesters can air their views and the minister be properly informed, but public inquiries in Britain are only advisory and not decision making. The minister can do as he pleases and there is no appeal. So once any nuclear proposal reaches the public inquiry stage the chances of failure are very small indeed. But is this sophisticated stream-rollering approach the best way to proceed in a democracy? How do you convince 'Joe-Public' that the 'experts' really do know what they are doing even if they have in the past failed to communicate their triumphs, skills, and concerns?

The nuclear industry still seems to think that there will continue to be little progress until there is a well informed public. There is talk about generations of believers being lost to the cause by their slowness to reach school children and tell them the truth about the nuclear industry. Is nuclearism a religious belief? Is believing in the sanctity of nuclear power similar to being a moonie? Some people clearly think so and this is very worrying. Of course most of the public will never be really well informed and thus perpetually unable to appreciate the scientific basis needed to give their complete and total support to nuclear technology. It *is* all really a matter of faith, and without blind and invincible faith (of the kind seen mainly in the nuclear industry) there are always going to be problems of public acceptability. However, this is only really a problem if (and only if) the aim of the exercise is 'brain-washing', which means persuading the public that the nuclear industry's version of the facts is the only correct version. There is a simpler strategy which suggests that a little give and take is all that is needed. It has been this reluctance to 'give' ground without any established scientific justification that has resulted in many of the problems and in accusations of political naïvety.

The answer is 'bend a little'. Be more sympathetic and adopt a long time horizon commensurate with the realities of the nuclear age (Openshaw, 1986). Whatever is being peddled as scientific truth today will probably be replaced within fairly short timescales, so why not be flexible, bend over backwards to gain long-term public acceptance, and regard 'science' as defining nothing more than a system of constraints within which public happiness criteria can be optimized. There is little mileage in seeking short term convenient solutions if the cumulative effect of continuing to be expedient instead of wise or intelligent is disaster.

If the nuclear industry's version of the 'truth' is non-absolute, ambivalent, uncertain and heavily assumption ridden, then herein lies a simple philosophy for moving Britain into a new publicly accceptable nuclear era. The answer is to err on the side of public concerns, to be explicitly sympathetic, to respond to seemingly irrational fears rather than attack their lack of science, to be humble rather than arrogant and always to plan ahead (Openshaw, 1988). The time scales in nuclear developments are never less than one century and may even be two. Who can tell what views/fears/concerns will be influencing the citizens of Britain in 2088 or 2188? Can anyone guess what amazing new scientific advances will have been made by then? More seriously, no doubt there will be much more detailed knowledge of low level radiation effects and probably far more stringent discharge limits and lower dose thresholds. No doubt also there will have been other Chernobyls in various parts of the world. The future is always uncertain, but what we now know with some certainty is that the current nuclear sites will still exist and that the radioactive waste dump being planned now will certainly be in place and maybe still in use. How can we ensure using current science that in 200 years the view will not be expressed that fundamental errors were made and that some of the scientifically reasonable 1980s assumptions used to prepare the safety case were not in error, and how can we be sure that the sites(s) will still be considered acceptable in a different era? Of course, we cannot be so sure. The prudent would deliberately seek additional margins of safety measured in an engineering–scientific fashion as much in terms of deferrence to public fears and concerns that may well increase rather than diminish with time. At risk is the whole nuclear infrastructure. The costs of closure and removal of a waste dump 100 years after it was first used might well be colossal but it might also be needed if there is insufficient lack of foresight now. Plan ahead, think ahead, eschew short-term expediency and play safe on a time scale measured in centuries rather than in parliaments.

An indication of traditional short-sightedness can be seen in the delayed recognition that nuclear waste was even a problem. This seemingly commonsense fact was only 'discovered' by the cognoscenti in the mid-1970s and it was not until the early 1980s that the problem seeped into public and media consciousness. So, some 30 years after the birth of the civilian 'nuclear industry' in the United Kingdom, and the creation of a substantial heap of radioactive waste, a major problem was discovered which is seen as a constraint on the civilian uses of nuclear energy. The 'Achilles heel' of all things nuclear seems set to become a visible threat to the glorious all nuclear future. Certainly the future viability of the nuclear industry does require that this waste problem be solved, not just in a theoretical engineering sense but with operational solutions in place. However, it can afford to wait longer than the antis may believe possible. Meanwhile, we are at risk because of the delays!

## Clumsy handling makes a difficult problem worse

The fact that nuclear waste is a difficult problem to handle, due mainly to political rather than engineering considerations, is hardly surprising given the nature of the stuff. Clumsy handling has allowed the anti-nuclear movements to become involved, to stir-up the fears, to secure the involvement of many previously apathetic non-politically minded people and to heighten media attention at all scales; no mean feat in such a short period of time. Yet the entire approach by the nuclear industry was doomed to fail at the outset. No government is going to risk losing many key seats on the nuclear waste issue which offers no benefits only problems. The analogy with nuclear power and its powerful institutional lobbies is wholly inappropriate. There has been a failure to approach the subject in a logical and scientific manner. Evidence of confusion and lack of coordination among the principal actors and a degree of secrecy and arrogance have merely inflamed public and media sensitivities. Further details of history are contained in chapters 2 – 4.

The least pressing problem of HLW was the first to be tackled in seemingly unnecessary haste – before there was any public debate, before any international experiences had been considered and before local communities were consulted. Then an attempt was made to force through a typical nuclear power station siting strategy for radwaste dumps that no community will probably ever want in their midst. As O'Riordan (1986) suggests, there was a failure to disentangle site appraisal from site selection, a failure to separate policy from route way and route way from site. The debate was moved from technical site-related issues to a political and strategic discussion of other aspects. What should have been a relatively straightforward task became a shambles of confusion and secrecy. As Blowers and Pepper (1987) point out, the result was a crisis in confidence that can now only be dealt with by more openness and frankness and with the acceptance of further constraints on plans for the future. A few years ago such thoughts would have amounted to heresy, now they are a small price to pay for acceptance.

It would seem that a nationally agreed waste strategy is clearly a good idea in principle, but extremely difficult in practice. The policy story is described in some detail in chapter 3. However, having a policy is one thing; it is not the policy aspects that are environmentally harmful and cause public despair, but their eventual concrete translation into one or more waste dumps built at particular locations. In particular, it is finding a site for these facilities that is perhaps the single most difficult task. There is likely to be little glory for those who try and solve this problem. Equally, should they get it wrong, then future generations may also revile them for their incompetence. If they get it right, then the public odium associated with the topic will ensure little reward for their efforts. Yet the problem is simply that of finding one or more sites for the safe disposal or indefinite storage of a cocktail of long lived radioactive bits and pieces. It seems simple enough. It is purely an exercise in applied geography; indeed, locational studies have always been fundamental and basic to geography. Yet despite this apparent simplicity, it is currently causing massive problems of all sorts in many different countries. It is unlikely that at the time of writing any country has satisfactorily 'solved' its nuclear waste disposal problems, and Britain is no exception. Some countries claim to have found acceptable disposal routes but it is too early to judge how successful they are likely to be. In this game the costs of failure can be high. You have to get it right at the first serious attempt.

## Handling the people

The question of public acceptability is now seen as of key importance; how it is to be overcome is a matter for debate.

The National Radiological Protection Board (NRPB) recommend a level of risk of death of 1 in 1,000,000 for a solid waste repository. Whilst this level is miniscule compared with most other hazards to life, such a statistic does not have much effect on public perception. It is apparent that the human view of risk is much more complex than a simple analysis of statistical probability. Fear of flying is far more common than fear of crossing the road, yet the latter carries a much greater risk. [Environment Committee, 1986, vol 1, p.xcvii].

This quote displays what in many ways is the heart of the problem. The nuclear industry seems to believe that if you can convince 'Joe-Public' of the truth of the 1 in 1,000,000 death risk statement then the problem is solved. The industry will demonstrate that their radwaste dump is at least this safe, the public will be happy that all is well and everything is OK. The problem is threefold: (1) the public do not understand the statement; (2) if they did, they would probably not believe it; and (3) their perception of risks is not based on objective facts and the existing 'facts' are based more on theoretical assumptions than on empirical reality.

Not surprisingly the nuclear industry believes that the main problem is public ignorance. RWMAC (1988b) claim that, 'The central need is to explain the risks from artificial sources compared to the larger risks from natural exposures to radiation and to the risks in other aspects of life and the natural environment' (p.43). They note, however, that the message was not getting through. Somehow a sense of perspective of the risks was being lost, and of the fact that the public had a very limited understanding of how radioactive waste is managed. The solution, according to RWMAC, is better targetting of the literature in the hope that once the key issues are explained, people can reach informed views. The problem here is will the people develop informed views that parallel those of the nuclear industry? Ignorance is undoubtedly part of the problem. O'Riordan (1986) suggests that 'unfamiliarity leading to dread' is a strong element in the public's anxiety but he also states, and we agree with him, that: 'This opposition cannot be countered by public relations tactics from within the industry, no matter how elaborate. Nor will advertising do the trick, nor even a barrage of "factual information" aimed at education and improving understanding' (p.529). The reason is simply that public anxiety is strong and deep-rooted and involves a very powerful fear of radioactivity, cancer and genetic disorders and also of the industry associated with it. Indeed, the Environment Committee (1986) comments that 'The key problem is the industry itself' (p.xcix), while O'Riordan (1986) stresses that 'radioactivity is associated with a technology and an industry that appears remote, self-confident, yet unaccountable. A substantial number of people are beginning to distrust "high science", and technologies that seem so complicated as to be beyond the capacity of elected representatives to understand them and therefore be in a proper position to make sound and informed judgements about them' (p.528).

There is unlikely to be any simple solution. The Environment Committee (1986) praise Lord Marshall when they write:

'Lord Marshall understood the emotional fears of the public and realised that the industry "has to be patient in explaining to the public that we believe we know what we are doing and the risks are modest", that "we would like to see steady progress but we do not see urgency attached to this matter". Our greater anxiety concerns the UKAEA and their perceived need to "demythologise" radioactive waste and BNFL's general attitude of frustration and irritation at ignorance stirred up by the "vocal minority" ' [p.cii].

The real mythology concerns the vocal minority argument. It might have been true once, maybe 10 – 20 years ago, but no longer. There is now unmistakable evidence that the vast majority of the country fear nuclear power and are opposed to further developments. The only prospect of reducing public opposition involves a combination of greater openness, greater accountability, bribery in the form of generous compensation for damage due to blight and as payment to a local community for accepting a radwaste dump in its midst, much greater public involvement, and also erring far on the side of safety in the cost-benefit analyses. The whinging about unnecessary expenditure on pollution control plant at Sellafield merely to save one or two lives really has to stop; for instance, BNFL (1985) said, 'We are spending 250 million pounds over the lifetime of this plant and by so doing we might avoid two cancers' (p.191, Environment Committee, Vol 2). It is a pity that BNFL had not then realized that they are paying not for two theoretical lives but for public acceptance. If the alternative is closure of Sellafield, then the 250 million pounds might well seem a good investment. The same strategy also has to characterize the radwaste dump if it is to succeed. The Environment Committee recognize this fact when they write:

... we conclude ... that a major cause of the public's extreme anxiety is not ignorance but distrust of the industry; that the industry must radically change its present attitudes and its relationship with the public; that a "Rolls-Royce" approach must be embraced and is not significantly more expensive than less cautious solutions. [p.xxxiii].

The authors agree and complement these suggestions by their own in chapter 8.

## Siting toxic waste dumps

Some evidence to justify the greater than expected public fears of nuclear waste can be seen by looking at comparable non-nuclear toxic waste. Many industries produce toxic non-nuclear waste as a by-product of their activities and this waste has to be disposed of, preferably somewhere safe although there is plenty of evidence that, historically, safe disposal was not a high priority. In any case whether 'safe' disposal actually happens in practice is difficult to determine and the relevant legislation is not always what it should be. Non-nuclear waste seemingly causes very little fuss despite the real and potential dangers associated with it. There are probably a large number of past and present toxic waste disposal sites and no one appears too concerned. Table 1.4 gives some comparative estimates of hazardous waste generation. The very large quantities of non-nuclear hazardous wastes being dumped in a poorly controlled manner should, all things being equal, be causing more concern. For example, how many people know that low level waste is also dumped at Dounreay as well as Ulnes Walton and Clifton Marsh? How many people know that there are 27 local authority landfill sites that are currently authorized for the disposal of long lived wastes? How many people know where the principal non-nuclear toxic waste dumps are located? Indeed lack of media attention to the handling of other toxic wastes effectively hides a large industry from both public and political gaze. The lack of public knowledge of where the non-nuclear toxic wastes are being dumped, and where the dumps are located, helps suppress fears and concerns. A low, at times invisible, profile seems to help and few organizations are sufficiently foolish to deliberately create problems for themselves by drawing attention to what they are doing. There are, it should be said, almost certainly real health risks involved with the disposal of many toxic wastes that are perceived to be innocuous. Some people are almost certainly ill, dead, or dying because of these sloppy waste disposal activities, although proof or blame is virtually impossible to establish. Furthermore, the absence of small area based national health and disease data prevents any detailed statistical associations being established similar to those being suggested for certain nuclear installations. It is not that these associations do not exist, it is merely that the necessary research work has yet to be performed outside a narrowly defined nuclear context. Indeed, doctors have long been speculating about the possible effects of toxic (and not yet defined as dangerous) chemicals and substances being dumped into the environment without any great concern or any thought to containment, but the data necessary to support the research needed to draw public and media attention to the problems do not yet exist. Additionally, the situation is massively complicated and there is no well organized anti- lobby on the prowl for evidence. So the problem remains 'hidden'. Even the chemical accidents at Seveso and Bhopal did nothing to emphasize the risks posed by the previously perceived to be safe non-nuclear industry. In terms of media impact and the longer term effects on people's perceptions of danger, these non-nuclear disasters are small compared with the effects of the reactor accident at Three Mile Island and the even more massive Chernobyl nuclear disaster which caused either no or very few short-term casualties. It is perhaps surprising that the nuclear industry has not made more political capital out of the great care they are required to give to safety compared with other industries.

*Table 1.4* Britain's annual toxic waste problem

| Type of Waste | Amount (million tonnes per year) |
|---|---|
| Hazardous industrial wastes | 10 |
| Coal spoil | 50 |
| Household refuse (England & Wales) | 28 |
| Radioactive wastes (estimated 2000AD) | |
| LLW | 1 |
| ILW | 0.16 |
| HLW | 0.004 |

Source: DOE (1985-6)

Contrast this situation regarding toxic chemical dumps with that facing the nuclear industry. Virtually every aspect of its operation is subjected to critical public gaze, all developments attract attention, it is the most heavily regulated of any industry, it is uniquely to blame for any and all radiation damages and injuries however caused, and any incident no matter how small and insignificant becomes immediate news and attracts massive media attention. The legacy of secrecy due to the largely historic association with nuclear weapons only serves to increase public suspicions. Indeed, now that a more 'open door' policy is being followed, the increased flow of information merely serves to increase public fears and concerns. In a fair world a far higher level of neurosis would also be associated with those conventional industries actively causing public harm without any recourse to pessimistic assumptions. Yet for reasons not difficult to understand, the nuclear industry attracts a far greater degree of fear and increasing intolerance than perhaps it deserves.

There is not going to be any easy solution to this problem and it is hardly surprising that the topic of nuclear waste disposal is of considerable public and thus political interest and concern. The facts of the situation, the nature of the problem, and the characteristics of the engineering alternatives are to some large degree probably quite irrelevant. The thought of living next to (or near) a nuclear power station is not viewed as being particularly welcome, especially since Chernobyl and the various cancer cluster scare stories that abound. In any case, how 'near' is 'near', and what is the critical danger distance? Neither can be defined in any sensible objective manner. No one denies that nuclear power reactors can be dangerous, merely that they are under control and there are no risks to anyone under all but the most pessimistic of circumstances; this is not the place to assess what is 'a reasonably pessimistic circumstance' or to inquire about the chances of any unreasonably pessimistic circumstances arising. If living near to a nuclear reactor is considered as being at least unwelcome, the thought of being a neighbour to a radioactive waste dump is even less tolerable. There is of course no great difference when measured in terms of total radioactivity and it is also clear that the reactor is a far less effective waste dump than a properly engineered structure would be. A working reactor contains in its above ground pressure vessel a vast amount of radioactivity and there is, by comparison with a properly designed radioactive waste dump, relatively little to stop it all escaping. However, no one is really impressed by such arguments because radioactive waste brings with it other problems and no direct benefits.

## What is special about the nuclear waste problem?

People seem to fear nuclear waste dumps even more than nuclear power facilities. Moreover, nuclear waste is often perceived to be vastly different from other types of toxic non-nuclear waste. For example, mercury is highly toxic and has a half-life of infinity. It does not decay with time but is merely redistributed by geomorphological processes. By comparison, plutonium-242 has a half life of only 387 thousand years and is not nearly as toxic! (See Table 2.2 for the half-lives of other radionuclides.) On the other hand, you have to ingest heavy metals before they have an effect. The consequences of Roman lead mining may still be visible in river terraces but they occasion little harm unless you eat sediments or grow cereals on river sediments. By comparison, had the Romans created plutonium 242 then the problem would be handled quite differently. So perhaps the basis of the comparison is wrong, in that infinite 'life' chalk is being compared with radioactive cheese and people perceive the risks as being quite different.

Some of the reasons for popular concern can be enumerated as: (1) there is no benefit in having a waste dump in your backyard; (2) the fears about health effects related to normal operation – 'leaks', accidents, and transport; (3) the long time scales involved making total containment impossible and causing ethical problems about leaving waste for future generations and even different civilizations; (4) monitoring and management will probably be required over extended periods; (5) the incremental nature of nuclear decision making may hide the eventual end state from public view until a later date when the development cannot in fact be stopped; (6) the effects on the economy and attractiveness of the surrounding region; (7) the traffic and other disturbances; (8) the increased risks from terrorist attacks and in time of war or civil strife; (9) a future dominated by media-inspired scare stories; (10) the fear of environmental damage; (11) the irreversible blighting and changes in the character and external perceptions of an area; and (12) the positive attraction of other nuclear industries into the area in the future.

## National interest arguments are critical to the debate

The principal problem seems to be that of public acceptability. This may seem strange because the nuclear industry was seemingly unconcerned about public acceptability. Historically, in siting nuclear power reactors the CEGB and SSEB always worked top down. You had merely to conform with the laws of the land, hold a public inquiry if required, and expect eventual success. This is called 'following established democratic procedures'. If you opposed the process, then in theory you were in contempt of the laws of the land. Local opinions and fears never entered into the process of decision making in a long established democracy such as Britain although there was some opportunity for airing views at the public inquiry. Provided that the necessary needs and safety case was well prepared, then eventual success was virtually guaranteed. This strategy worked well during the early years of civilian power, it worked well during the Windscale inquiry, and it still worked through the lengthy Sizewell inquiry. Costs were measured in terms of money and delays. At the end of the day, national considerations usually over-ruled local objections. The justification is simply that a developed country needs electricity and this national good more than compensates for local damage.

Certainly, as chapter 4 shows, the early efforts to solve the waste disposal problem followed this traditional pattern. The so-called CEGB approach to siting (as detailed previously) was adopted and the steam-rollering process started. Yet twice, it seems political arguments intervened. Now it appears that a 'let the people decide' strategy is being adopted as the best way out. Whether the elitist scientific arguments that normally characterize the siting process can be so easily displaced is too soon to say, but it does mean that a new found degree of pragmatism will probably favour the easy to develop and the relatively non-controversial site, rather than a decision based purely on engineering concerns. Let us hope that it is not at the expense of technical excellence.

However, it is also clear that the 'National Interest' argument will still be used in order to justify even a relatively non-controversial location. If this is the case, then it seems essential to know: where are the sites that could qualify as nuclear waste dumps? How many sites are there? How does site X compare with site Y? This sort of justification seems essential to sustain any national interest argument: namely there should be some assurance that the best location or nearly best location has been found based on technical excellence and suitability criteria, rather than just any old site that happened to be available and convenient. It is also necessary for planning purposes. There can be no assurance that only one major radwaste dump is all that will be required, although only one site is currently being talked about. However, the future needs of the industry may be such that others may soon be needed but no doubt this 'discovery' will only be made after the first decision to proceed is taken. This is the classic incremental decision making process that serves as long term planning in the nuclear industry. It offers the advantages of flexibility and convenience, it also tends to separate developments in time and space and thus by a process of disassociation to segment and thereby minimize the total effective opposition that might otherwise arise. Again it would be very handy to know more or less 'where' the potentially suitable locations are now, so that the debate for the first site (wherever it may be) can be properly informed about the *total* picture, rather than just one very small, site specific part of it.

A central theme throughout this book is that recent developments in Geographical Information Systems (GIS) allow a much more detailed and automated preliminary site search process than was previously possible. Manual map overlaying methods can be replaced by computer based GIS equivalents which offer significantly greater search and evaluation possibilities. We can now seek to identify the set of all technically feasible sites and then look at the relative performances of any preferred locations. We fear that previous siting attempts have been hindered by a failure to use this new computer technology to the maximum effect. Suddenly, the national interest criteria becomes more difficult to satisfy, and quite rightly too. The burden of proof should now pass to the developer. Let them prove in public that they have found the best site(s) in the national interest and not merely industry optimal sites that are very poorly located and far from being near the top of the site charts. This theme is developed further in chapters 7 and 8.

### A double-edged sword

On the other hand, perhaps it is all too easy to exaggerate the importance of a problem that is really not a problem at all. The so-called radwaste dump problem is only perceived to be a problem by the nuclear industry because of the difficulty of obtaining approval for a facility. It is a problem from the public's point of view because they probably do not trust what the industry tells them about it and because post-Chernobyl anything nuclear is fearsome. It is a problem from a governmental point of view because something has to be done and whatever is done will lose votes. It is one thing pushing ahead with nuclear power developments which are only modestly unpopular and which can be proven to be beneficial, it is quite another proposing guaranteed vote losing radwaste dumps which are only very indirectly beneficial and bring with them a blighting factor of considerable psychological proportions compared with the actual physical impact. However, perhaps none of these are really 'problems' at all! Perhaps all the parties concerned are creating a problem by treating a task as a problem and then, because it is perceived to be a problem, being afraid of it, making it one thousand times worse that it really is (or was). Perhaps even this book, by drawing attention to the subject as a problem, is merely making a non-problem worse.

Maybe, for the next 50 years, it would be economical and acceptable simply to store rather than dispose of the various wastes while waiting for the disposal technology to mature and for public attitudes to change. This would certainly be feasible and represent a quick and easy solution. Yet it is currently ruled out for 'policy reasons'. Ever since 1982, the government's policy has been to dispose rather than store wastes and, because of the surprising reluctance of the 'independent' and 'expert' advisers to recommend any other course of action, it has effectively created the radwaste dump problem. This reluctance to contemplate storage, something which is seemingly viewed as an unneccessary and suboptimal course of action, reflects the engineer's view of the world. If the technology exists and the only problem is people, then there is no problem! The reality of the last few years has proven otherwise. Yet it could all be avoided, probably even now, by seeking storage until public acceptability is no longer an issue. Or is the storage option not really acceptable? Maybe it is too expensive, maybe there are doubts about packaging, maybe there are doubts about the integrity of the engineering designs, maybe they do not really know what they are doing or properly understand the processes involved. Dumping out of sight avoids all these problems and leaves future generations the task of determining how good the designs really were.

This chapter is deliberately written in what might be described as an emotional style. Group therapy is a good way of exorcising pent-up fears and tensions at the earliest possible moment. It is not possible to write and research about such emotive issues as radwaste without being affected. It is also essential to recognize that there are very different and often quite incompatible perspectives of the radwaste problem. Indeed there are at least five different public perspectives that are of interest here.

View 1 is purely and totally scientific. Judgements and actions are supposedly based on the best available scientific advice in the context of a positivist philosophy of the world. Decisions are seen as being neutral, value free and based on hard, objective information. Facts can be established and the 'correct' decisions made. No degree of explanation will ever convince a hardline scientist of the need or appropriateness of making and taking decisions based on non-scientific grounds or, perhaps, of the reality of intense and irrational fears based on what appears to be no more than ignorance. This belief in the supremacy of science often induces elistist feelings and sometimes a degree of arrogance.

View 2 is the other extreme. It is purely emotional rather than scientifically based, but is no less real or narrow minded. The thought of living near to manmade radiation created from a commercial activity that gives off invisible death rays, threatening all about it, and slowly injuring and killing children and destroying nature can be simultaneously nauseating, terrifying and so utterly awful that virtually anything is justifiable to avoid it coming 'here'. No degree of explanation or education will ever persuade some people of the irrationality and unreasonableness of their fears. Radiation is fearsome and totally dreadful and is likely to kill and injure many innocent people because it is neither fully understood or under control. It is easy to convert people to this way of thinking.

View 3 is based on a legalistic and political version of the truth and is perhaps more cynical. Absolute truth or emotion no longer matter. The name of the game is balancing probabilities, the manipulation of the facts, balanced judgements made on behalf of others for personal or party or client gain, and an endless search for political advantage. Truth and facts, experts and public are mere distractions; artefacts to be used when useful and discarded when not.

View 4 is probably that shared by the man in the street. These people are not scientists, they are not interested in litigation, nor are they as emotionally involved as some anti-nukes, but ordinary people. They tend to believe what they read in the popular press. Radwaste is best avoided for no good reason other than it is labelled as potentially dangerous, harmful or nasty. What could be worse than having it near where you live. No amount of education is probably ever going to entirely dispel their fears; once they exist they will linger on. There are some exceptions, mainly nuclear industry employees or others who see some benefits in the developments. Many would regard it as unnecessary while there are non-nuclear alternative sources of power. Each new scare story tends to strengthen their beliefs in a vague sort of way. The public relations damage is cumulative but the intensity has a short half-life!

Finally, view 5 is that being promulgated by the authors. It is argued that to make any impact it is necessary to be able to understand the feelings and views of all four groups. It is necessary to compete with the scientists on their territory by introducing methods and advice which are scientifically justifiable but which are also seen as moving science to accommodate the feelings and fears of the man in the street. This book is concerned with trying to help both those who want to find a solution and, also, those keen to oppose any solution. It attempts to de-mystify the nuclear waste problem, to socialize the science, to provide an independent and critical review of what has been going on during the last decade or so, and to make some positive suggestions based on GIS technology as to how the problem might best be handled in the national interest. Without wishing to be overly melodramatic, it is important that the radwaste problem is overcome in a widely acceptable fashion. Put quite simply, if no satisfactory and publicly acceptable solution can be found, then the continuing production of waste and consequently the ultimate future viability of nuclear power itself would be threatened. This situation will of course never occur because no government could possibly allow it to. All this because of the difficulties of siting Britain's radwaste. It would be far, far better to resolves the issues sensibly, balancing rational science with the need for public acceptability.

# 2

# A review of Britain's radioactive waste problem

### Radwastes and the nuclear fuel cycle

Let us look again at where Britain's nuclear wastes come from. Some historical detail is also useful in order to understand how the present situation has been reached and what is likely to happen in the future. The waste stocks described in Table 1.1 are believed to represent most of what is left after more than 40 years of atomic experimentation, weapons development, fuel reprocessing, power generation and dumping activities. The vast majority is widely regarded as being due to the reprocessing of nuclear fuel, although this is not the impression that is often given: see for example, Table 1.2. Originally, this reprocessing activity was to acquire plutonium for nuclear weapons but it is now mainly viewed as a waste management technique which is part of the nuclear fuel cycle.

*Table 2.1* Sources of total projected nuclear wastes (m³)

| Type of waste | 2000AD total | 2000AD decom | 2030AD total | 2030AD decom | post 2030AD decom |
|---|---|---|---|---|---|
| Low level | 423,560 | – | 1,411,000 | 250,000 | 716,000 |
| Intermediate | 99,770 | – | 259,200 | 65,000 | 203,100 |
| High | 1,193 | – | 3,030 | – | – |

Note: decom = decommissioning wastes
Source: RWMAC (1988b) pp.53, 55

Table 2.1 gives estimates of future wastes that will need disposal by the three major categories that are usually recognized. It should be noted that these waste streams are not homogenous and they represent a cocktail of many different radionuclides with a wide range of half-lives; see below. Note also the increasing contribution from decommissioning: indeed by 2030AD these wastes may well equal those expected from reprocessing activities. A large part of these future wastes are unavoidable because they reflect the consequences of nuclear power and reprocessing decisions that have already been made. Their radwaste implications merely lag behind by 30 – 100 years. It is, however, a fact that the amount of radioactive waste thought to exist is a variable quantity. It depends on the comprehensiveness of the waste inventory, the nature of the packaging assumed to be used and estimates of the levels of compaction obtained. Additionally, the quantities of decommissioning wastes are unpredictable and depend on the nature and timescale of the decommissioning process, while the estimated reprocessing arisings take into account decisions about new reprocessing plant that will not be made until 2010AD or so. The quantities of radwaste are therefore technology dependent although this really only influences the margins (perhaps 10 to 15 per cent of the totals). The figures in Table 2.1 include both 'committed' and uncommitted wastes. The former reflect decisions already made and the later estimated effects (in 1988) of the radwaste consequences of decisions not yet made. The numbers refer to what are termed 'conditioned' wastes as distinct from 'raw' wastes. These are used because they provide a better indication of possible dump space needs. Finally, the low level wastes figures refer to those wastes that leave the site where they are created; this excludes the dumping activity at Drigg. All the figures also exclude the activities of the MOD.

It should be noted that reprocessing serves both a military and a civilian purpose. All the early efforts were designed to acquire plutonium for nuclear weapons. The Windscale reprocessing plant (1951) and the Windscale Piles (1951–7) were explicitly designed for this purpose. The subsequent reprocessing of spent fuel from civilian MAGNOX reactors was also initially to meet military needs although it is unclear as to whether any, or much, of these stocks were actually used for weapons manufacture. Of the wastes listed in Tables 1.1 and 2.1, it is also not known what proportions resulted from the reprocessing of fuel to extract plutonium for nuclear warheads. All statistics on radioactive wastes from defence establishments are restricted for security reasons (Cmnd 8607, 1982; para 14). It is likely however that the exclusively military wastes are 'lost' in the civilian waste heap. Current MOD arisings are thought to be no more than 20 per cent of civilian wastes.

Alternatively, it is possible that the real reason for secrecy is that the majority of the wastes from the early weapons development era were dumped either in the Atlantic or the Irish Sea; or even on land. Official Secrecy (and probably also accounting problems) ensure that the real answer may never be known; indeed, does anyone care? It is noted that an important military interest undoubtedly continues; the reactors at Calder Hall and Chapelcross are not primarily there to generate electricity and there has been a long debate about the extent to which plutonium from Britain's civilian power reactors may have been diverted to weapons development. The co-processing of both civilian and military irradiated fuel at Sellafield makes plutonium accountancy suspect because diversion is theoretically possible. On the other hand, it is also quite obvious that the main enthusiasm for reprocessing is now not weapons-related but is connected with preparations for the development of fast breeder reactors. It is also important to note that Sellafield is the largest source of all the wastes; in 2000AD this is estimated to be: 87 per cent of the high level wastes, 69 per cent of the intermediate level wastes, and 59 per cent of the low level wastes (RWMAC, 1988b, p.53). These wastes are largely due to the reprocessing of nuclear fuel to recover both uranium and plutonium from the spent fuel and to provide a safe way of handling spent fuel by concentrating the radioactivity. It is useful then to have a closer look at the role of reprocessing in the nuclear fuel cycle.

The fuel used in reactors passes through a cycle which consists of three stages: the front-end, reactor fuel, and the back-end. The front-end starts with the mining of uranium and includes its subsequent processing and enrichment, fuel fabrication and its arrival at a reactor. The second stage begins when the fuel is irradiated inside a reactor. After a while it is removed, stored on site to cool and then transported to a reprocessing or storage facility. The third stage, the back-end of the cycle, is the 'dirty' end. Here the spent fuel is reprocessed to recover the reusable uranium and plutonium and separate out the wastes for storage or disposal. It is noted that nuclear power is fairly pollution free because nearly all the potential pollution nasties are held within the spent fuel and then subsequently concentrated during reprocessing. This three stage process is termed a cycle because both the uranium and the plutonium can be re-used as reactor fuel. Reprocessing recovers about 99.9 per cent of the plutonium that was created by the fission process and virtually 100 per cent of the uranium that was in the original fuel. The typical composition of irradiated MAGNOX fuel is 99.2 per cent uranium, 0.3 per cent plutonium, and 0.5 per cent waste fractions (CEGB, 1985, p.85, vol 2). Clearly, it is a highly efficient process.

Recycling the separated uranium can improve the utilization of uranium by about 20 per cent thereby reducing the total demand for uranium. Indeed by 1983, about 80 per cent of the AGR fuel was based on uranium recovered from MAGNOX reprocessing, although this is no longer the case. Likewise a thermal recycle of the plutonium can add a further 15 per cent to the energy recovered. However, if the recovered uranium and plutonium is recycled in a fast reactor then the total energy yield from the fuel can be increased by a factor of 60 (BNFL, 1985b). The government's view mirrors that of the nuclear industry; namely, 'only by reprocessing the spent fuel ... do we preserve the opportunity to utilise the uranium and plutonuium, which represent a valuable energy resource equivalent to the entire commercially recoverable resources of coal in the UK' (DOE, 1986a, para. 76). This is an extremely powerful argument of considerable long term significance. Reprocessing is seen therefore as the key to both the conservation of uranium resources and security of supply in the long term. Its economic viability, like the fast breeder reactor, is directly related to the price of uranium and the value attributed to the plutonium. Currently there are no signs of any impending shortage of uranium but taking a long term view then it is not difficult to envisage shortages developing in the increasingly all-nuclear world of the future (maybe in the 22nd century or later). So there is a powerful argument that it makes good sense to conserve uranium and reprocess so that the fast reactor option can be kept open for when it becomes necessary. This view is well established within the nuclear industry and no doubt it will continue to prevail. The arguments in its favour include: the need to fill the energy gap left by declining oil and gas reserves, the need to be independent of world fuel and energy prices which can be manipulated for political ends, the need to be independent of imported uranium, and the need to generate cheap electricity in a safe, environmentally clean, and reliable manner. The FBR is viewed as having the potential to meet all these requirements (Bloomfield, 1988).

The key question concerns the nearness of the FBR era. If it lies 50 – 100 years into the future, then there is probably little real justification for reprocessing rather than storage or direct disposal of spent fuel. If however, as it has often been perceived to be, it is only five to ten years away, then there is every reason to develop reprocessing technology now because of its essential role in the FBR fuel cycle. What has happened is that for the last 30 years or so all the forecasts have been far too optimistic. In 1959, the House of Commons was reliably informed that the first commercial FBR would start in about 1970. As late as 1976, the United Kingdom Atomic Energy Authority (UKAEA) told the Royal Commission on Environmental Pollution that it envisaged some 33GW (equivalent to about 27 AGR stations) of FBR being in place by 2000. Too many nuclear decision makers have had far too much uncritical faith in the accuracy of their own forecasts of the future. This constantly wrong forecasting of when the commercial FBR era will start has hindered the formation of a rational nuclear policy for at least the last 20 years. It also appears to have encouraged the seemingly unnecessary expenditure of large amounts of nuclear research and development on fast breeder technology which is unlikely to be commercially viable in the foreseeable future. Indeed the recent CEGB intention to withdraw its support for FBR developments reflects a realization that it is still, despite major advances in recent years, far from the market place. The implications are that a few more generations of PWR reactors are likely before a commercial FBR becomes realistic. On the other hand, European interest in commercial FBR demonstrator projects certainly still seems alive and well. Yet it should not be forgotten that in the 1960s there was strong support for 'commercial' FBRs which if built would today have been largely uneconomic by comparison with the best alternatives, although maybe much less so against the worst. Nevertheless, this hidden sustained support for what is still an uneconomic means of power generation is a classic example of how in the nuclear field a powerful, invisible, extra-democratic establishment lobby has been able successfully to channel basic science research monies into a pet project. Once started down this route, sheer inertia (particularly the costs of closure) tends to keep it running.

These FBR developments and dreams continue without any proper realisation of the potentially tremendous political implications that dependency on a plutonium economy might well entail. The concern over the risks to civil liberties should not be dismissed too lightly. The difficulties associated with the requirement to ship possibly large quantities of plutonium with zero risk of diversion, theft or environmental contamination should not be underestimated, especially at a time of increasing international terrorism and, no doubt before too long, atomic weapons availability to potentially unstable political regimes. To some extent, the same type of societal problems that are linked with the possible plutonium economy of the future can be seen in the radwaste issue. The problems that remain are as much political and social as they are engineering challenges. The recent industry frustrations over radioactive waste management are a classic example of what happens when engineers neglect the wider socio-political context and the long term societal implications of their pet projects. Nothing to do with nuclear radiation can today be regarded or handled in either a purely rational or objective scientific manner. The 'scientific facts' are incidental and largely irrelevant to public perceptions of fear and concern; and it is unlikely that this attitude of mind can be easily or speedily changed. The history of Britain's attempts to solve the radwaste disposal problem clearly illustrates the results when politically naïve technologists ignore this broader non-scientific environment in which, whether they, like it or not, have to operate (Openshaw, 1988). The end result will probably be appeasement via a deliberate and potentially expensive abandonment of industry optimal solutions in the direction of improved public acceptability. It will cost more to develop these scientifically over sophisticated, so-called 'Rolls-Royce solutions'. However, the alternative is probably no solution at all and a political backlash with a gradual erosion of all things nuclear to the probable ultimate detriment of everyone.

## Reprocessing as a waste management tool

In Britain, nuclear fuel reprocessing is performed at Sellafield and Dounreay. Dounreay reprocesses its fast reactor fuel on an experimental basis. There was a planning application in 1986–7 to establish a European FBR fuel reprocessing facility there. The development of this facility is currently in doubt because FBR costs are still uneconomic and do not seem likely to rival PWR generations until 2000AD or even much later. Sellafield started reprocessing in 1952 and was taken over by BNFL in 1971 from the Ministry of Supply and the UKAEA. In excess of 25,000 tons of MAGNOX fuel has now been reprocessed. Sellafield is where nearly all Britain's civilian and military nuclear fuel has been and is being reprocessed.

The idea of a nuclear fuel cycle which neatly dovetails thermal reactors, fast reactors, reprocessing, and waste disposal into a single linked process is very appealing. It has both an innate neatness and a compelling logic to support it. It is also strategically very important in the long term but is it currently essential?

Clearly, the nuclear industry thinks it is and have invested quite large sums of money in their faith. Some other people seem to think otherwise. Yet to a large degree reprocessing is not optional in Britain but virtually mandatory. Quite simply, it is an unavoidable corollary to having MAGNOX power stations. They are likely to last another 10 years and so will the need to reprocess their fuel. The reason is that as soon as spent MAGNOX fuel goes into water for cooling it starts to corrode and must be reprocessed within two or three years. MAGNOX fuel could have been cooled in a dry store but only one of the MAGNOX stations is fitted for dry storage. Back-fitting the remainder is not considered sensible because of the cost, their short remaining life and the low density of MAGNOX fuel; moreover, the fuel may still have to be reprocessed. Quite simply, the MAGNOX stations were built with reprocessing in mind. In the 1950s and 1960s it did not occur to anyone not to reprocess; indeed, an important factor in building MAGNOX was seemingly to increase military plutonium stocks by reprocessing, although this option does not seem to have been used. Spent AGR fuel is also stored under water awaiting a new thermal oxide reprocessing plant (THORP) to come into operation at Sellafield and none of the AGR fuel has been reprocessed. Yet once AGR fuel goes under water, it too will probably have to be reprocessed albeit on a much more relaxed time scale. Currently, all the spent AGR fuel is stored under water at Sellafield. Only the PWR fuel (of which there is not any indigenous civilian British supply as yet) could be stored more or less indefinately under water until such time as reprocessing or disposal is selected Indeed, in the United States there is no civilian fuel reprocessing and the spent fuel is simply stored. A critical aspect here seems to have been a lack of interest in the need for fast breeder reactor technology. Uranium supplies have always been much less secure in Europe than North America, thereby making the FBR an insurance policy against either shortages of uranium or political manipulation of uranium supplies at some future date. These arguments appear today to be very weak, since it would be much cheaper simply to stockpile uranium especially as the price is currently so low.

The reprocessing versus storage situation is further complicated by economic arguments. At present it probably makes more sense to store PWR fuel for reprocessing when this becomes economic or for disposal at a later date. AGR fuel can be stored, probably indefinately, in dry stores and there are plans to construct one as a buffer storage alternative to Sellafield. Storage gives greater flexibility in the short term but it also occasions additional costs when there is a reprocessing option. At present it seems that the costs of storage and reprocessing are essentially similar. Storage could be indefinate, but it is probably only reasonable to insist on disposal or reprocessing in the longer term. Moreover, both the MAGNOX and AGR fuel may well have to be reprocessed.

There is also some concern about leaving spent fuel, which is essentially a HLW source, in surface storage, where it is accessible to the environment and requires constant monitoring, active care and supervision to ensure safety.

Storage is usually considered to require a level of institutional support that might not be realistic on an extended timescale: for instance, 200 years is usually taken as a reasonable limit. Why 200 years is acceptable but 500 years is not, has not really been adequately explained. Certainly, life 200 years ago was very different from today! There is also an argument that many aspects of the human legacy imply long term institutional support, for example, dams. Why single out nuclear waste as being different? If nuclear war wipes out most of the planet in 50 years, surveillance of radwaste dumps will be of trivial significance. Hence, the strange almost illogical fear of a greater than 200 year institutional responsibility. If we are all dead, does it matter?

Further complications arise when nasty questions are asked such as: is reprocessing necessary as a step towards final disposal or is it merely a means of generating income for BNFL? Why are so many countries so keen to pay high prices to have their fuel reprocessed in THORP when it would be considerably cheaper to buy new uranium supplies? Why do the CEGB regard their spent fuel as a resource rather than as high level waste? One answer is that reprocessing solves the problem of what to do with spent fuel; in Britain the solution is send it to Sellafield and have it reprocessed. Another pro argument is that reprocessing is currently possible. Tighter environmental controls or problems due to the corrosion in dry storage of fuel, may conceivably preclude this activity in the future, so why not make good use of the existing opportunities! There are also political arguments that might imply a British attempt to dominate the world reprocessing market as a force against nuclear weapons proliferation by discouraging Third World countries from starting to develop their own facilities. This all seems rather far fetched and it assumes a degree of foresight that is not usually evident.

The truth of the matter is probably much more straightforward. In the mid-1950s when the nuclear age was being planned no one considered that MAGNOX fuel would not be reprocessed. Having gone to the expense of building reprocessing plant, it might as well be used. The AGR fuel was also intended to be reprocessed. So after considerable delay THORP is being built. Having decided to build THORP it makes no sense to spend over a billion pounds on a facility that is not going to be fully used, but why build one three times bigger than home demand could seemingly sustain? Is THORP another example of a large scale planning disaster because of widely inaccurate forecasting? Or is it there because BNFL was keen to expand reprocessing as an essential element in their commercial strategy? The latter seems to reflect what in fact has happened. Reprocessing is government policy but the size of the reprocessing programme was left to BNFL. Indeed, BNFL (1985) comment: 'Spent fuel management policy is not simply a question of technical and economic argument but is bound up with sensitive commercial, strategic, and political value judgements' (Environment Committee, 1986, p.xcv, vol 1). It is a pity then that no one seems to know what precisely these 'sensitive commercial, strategic, and political value judgements' are all about.

It appear that there are other good reasons for wishing to reprocess rather than store spent fuel. They include: (1) leaving plutonium in spent fuel is potentially harmful; (2) reprocessing concentrates the wastes – the high level wastes contain some 95 per cent of the radioactivity; (3) vitrified high level wastes may be more leach resistant than spent fuel; (4) disposal without reprocessing denies access to valuable fuel resources; and (5) the facilities exist and enthusiasm for a nuclear fuel cycle remains strong at all levels where the key nuclear decisions are likely to be made.

The whole reprocessing business would sound so much more convincing and worthwhile if the plutonium had some value, currently it is a worthless economic commodity. Any potential value it may possess is either for bomb making (if it is good enough) or when it becomes an economic fuel at some indefinite future date. Nevertheless, since 1969, the plutonium produced by the reprocessing of irradiated fuel from civil reactors has been owned by the Central Electricity Generating Board (CEGB). Prior to that time it was the property of the UK Atomic Energy Authority (UKAEA) and thus of the government. The CEGB's policy is to use its plutonium as a fuel for a possible future programme of fast reactors or as a fuel in the thermal reactors of the future. In 1978 the United Kingdom voluntarily accepted International Atomic Energy Authority (IAEA) safeguards on its civilian power reactors, the aim of which is to prevent the diversion of nuclear materials for military uses. The size of the plutonium stockpiles is still a national secret. If it is mainly intended as FBR fuel, then the question arises as to whether it is economic to invest so heavily in the future when by so doing such large quantities of radioactive wastes are being generated. The answer is that an element of choice probably only really exists for future PWR fuel. So although there are some fairly compelling reasons for wishing to halt reprocessing activities, because it is by far the largest source of radioactive wastes and therefore to a large degree determines the size of Britain's radwaste problem, there is no real option at present. It is also useful to remember the growing contribution to the future radwaste heap that will come from decommissioning and the smaller arisings from nuclear power generations, nuclear weapons, and research activities. It seems that the radwaste problem is here and will not easily go away either now or in the foreseeable future. It is also worth noting that reprocessing as such does not create radioactivity even if it does increase the volume of contaminated materials by about 200 times. It can be justified as a waste management tool because it concentrates about 99% of the radioactivity into the HLW which is regarded as a more secure form than leaving it in the original spent fuel.

It follows therefore that the magnitude of the future radioactive waste disposal problem is largely a function of the amount of reprocessing that will be performed and the effects of nuclear decisions that have already been made and cannot be reversed. It is uncertain as to whether THORP is an option that can now be cancelled because of the large stocks of spent AGR fuel in storage that may well need to be reprocessed. So it is likely that whether we like it or not there is probably going to be a considerable amount of reprocessing performed during the next 20 years or so, not least because the £1,400 million cost of

THORP has been largely paid for by advance orders (mainly from overseas) and it is difficult now to see how this activity can be stopped in the short term even if there was the political will to do so. On the other hand, the CEGB are planning a £200 million irradiated fuel store at Heysham to avoid the future need for reprocessing of at least some of the spent AGR and probably all the PWR fuel. This allows the CEGB to strengthen their price bargaining position (it is currently much cheaper not to reprocess fuel) and also to provide buffer storage in the event of reprocessing becoming unavailable (due either to accidents or future political interference at THORP). A privatized nuclear power industry might well regard reprocessing as an expensive option at present.

It is interesting to note once more the uniquely long timescales that are associated with many nuclear activities. For instance, the current AGR irradiated fuel stored under water at Sellafield awaiting THORP reprocessing can probably only be safely stored for about 15 years. The development of a dry store at Heysham would probably increase the storage period to 50 years or more. The Sizewell B PWR fuel will probably be reprocessed by the successor to THORP operating in the period 2030–2040. The construction of each new nuclear power station carries with it considerable, albeit delayed, radwaste disposal implications on timescales of over 100 years (or more). This also means that there is certainly no need to start panicking just yet.

## Types of waste

It is current practice to classify radioactive wastes into three streams: high-level wastes (HLW), intermediate level wastes (ILW), and low level wastes (LLW). The meaning of these categories has only recently been standardized. Nevertheless, each is seen as presenting a different potential level of hazard and requiring varying forms of treatment and handling. Some of the wastes are intensely radioactive and are also heat generating (as the fission process continues), others require no shielding but are a hazard for millions of years, others could be dumped virtually anywhere without any control or concern. In practice all types have always been regulated by licences and authorizations. These controls have also become increasingly stringent as public interest (mainly fears about possible but not yet substantiated health impacts) have increased. It is useful then to describe in more detail these different types of waste.

*High level wastes (HLW)*

The smallest quantities of waste in Table 2.1 are also the most radioactive. These HLW, also sometimes referred to as heat-generating wastes (HGW), require that the heat produced by radioactive decay is taken into account in the design of storage or disposal facilities. The HLWs contain a large proportion of all the radioactivity that was present in irradiated spent fuel rods; something like

97 per cent is concentrated in these wastes. There are currently about 1,200 cubic metres of these HLW wastes resulting from MAGNOX reprocessing. The wastes are stored at Sellafield in high-integrity stainless steel tanks fitted with cooling coils to remove the decay heat from the fission products. The long term intention is to dry the wastes, trap them in liquid glass, and then store the vitrified product in stainless steel containers. These containers will then be air cooled. It is planned that these blocks will remain in storage for at least 50 years after which time the smaller amount of decay heat will make disposal easier. However, it is noted that some of the constituents of the HLWs have very long half-lives. Table 2.2 gives half-life details of some of the isotopes present in radioactive wastes. A half-life is time taken for the activity of a specified radionuclide to reduce by a half through radioactive decay. It may be assumed to be harmless when it has decayed for 10 half-lives. Another way of expressing the longevity of the potential danger is to compute the times needed for the rate of emission (ie radioactivity) to decay to 1/1000th of the original levels. Estimates are shown in Table 2.2. Note the extremely long time periods that are sometimes needed to achieve this and also the extremely long half-lives of some of the radionuclides. Beware! No indication is given of the relative quantities of the various radionuclides in a typical radwaste dump. It should not be assumed that long-lived radionuclides are always present and that they dominate. Nor should it be assumed that people know precisely what is in current waste dumps or that 'typical' is a term that can be applied here.

Some of the half-lives are so long, it is likely that even the most successful storage facility will be unable to prevent their eventual escape; the best that can be hoped for is that their return into the environment is greatly delayed. The case for disposal or continued storage is difficult on moral and ethical grounds. Has the current 20th century civilization any right to leave such a horrible mess for subsequent generations to handle and suffer any effects due to leakages etc? On the other hand, these HLW tanks constitute an international hazard of unparalleled magnitude simply by having such highly active liquid wastes in a form where they could be readily (in the sense of feasibly) dispersed (by fire or explosion due to accidents, terrorism or war) and contaminate vast areas with immensely long-lived radionuclides. If anything, these risks are the greatest of any associated with the nuclear waste problem. Maybe there is a good case (from a public and pan-European safety point of view) to deposit these wastes deep underground as soon as possible. Continued near surface storage must be viewed with some concern because of their theoretical vulnerability to external events beyond anyone's control. A major fire in a liquid HLW storage tank at Sellafield would make Chernobyl look like a 'teddy bear's picnic'. RWMAC (1988a) observe that the currently proposed date for a HLW dump is 2040AD. They note with more than the usual degree of cynicism about government policy that: 'We were pleased to see that DOE were keeping in touch with relevant generic studies and that the next significant stage of HGW research would be site specific studies that could be fitted conveniently into a 5 year period starting in about 2010' (p.3).

*Table 2.2* Times taken for various radionuclides to decay to 1/1000th of their original radioactivity

| Nuclide | Half-life | Number of years to decay to 1/1000th original level |
|---|---|---|
| Xenon-133 | 5.2 d | 0.0658 |
| Argon-41 | 1.8 h | 0.0658 |
| Radon-222 | 3.8 d | 0.1038 |
| Iodine-131 | 8 d | 0.2185 |
| Cerium-141 | 33 d | 0.9012 |
| Niobium-95 | 35 d | 0.9558 |
| Strontium-89 | 51 d | 1.3 |
| Yttrium-91 | 59 d | 1.6 |
| Zirconium-95 | 64 d | 1.7 |
| Cobalt-58 | 72 d | 1.9 |
| Curium-242 | 163 d | 4.4 |
| Cerium-144 | 285 d | 7.7 |
| Ruthenium-106 | 1 y | 9.9 |
| Promethium-147 | 2.6 y | 25.9 |
| Cobalt-60 | 5.3 y | 52.8 |
| Krypton-85 | 10.8 y | 107.6 |
| Tritium | 12.3 y | 122.6 |
| Plutonium-241 | 15 y | 149.5 |
| Europium-154 | 16 y | 159.4 |
| Curium-244 | 18 y | 179.4 |
| Lead-210 | 22 y | 219.2 |
| Strontium-90 | 28 y | 279.1 |
| Caesium-137 | 30 y | 299.0 |
| Plutonium-238 | 87 y | 867.2 |
| Samarium-151 | 93 y | 927.0 |
| Nickel-63 | 120 y | 1,196.1 |
| Americium-241 | 433 y | 4,316.1 |
| Radium-226 | 1,600 y | 15,948.6 |
| Carbon-14 | 5,700 y | 56,817.0 |
| Plutonium-240 | 6,600 y | 65,788.1 |
| Thorium-229 | 7,300 y | 72,765.6 |
| Americium-243 | 7,370 y | 73,463.4 |
| Plutonium-239 | 24,000 y | 239,229.6 |
| Thorium-230 | 80,000 y | 797,432.0 |
| Technetium-99 | 210,000 y | 2,093,259.1 |
| Uranium-234 | 240,000 y | 2,392,296.1 |
| Plutonium-242 | 387,000 y | 3,857,577.5 |
| Zirconium-93 | 1,500,000 y | 14,951,850.8 |
| Caesium-135 | 2,000,000 y | 19,935,801.1 |
| Neptunium-237 | 2,100,000 y | 20,932,591.1 |
| Iodine-129 | 17,000,000 y | 169,454,309.6 |
| Uranium-235 | 710,000,000 y | 7,077,209,405.0 |
| Uranium-238 | 4,500,000,000 y | 44,855,552,567.0 |

Notes: y = years, d = days, h = hours

*Intermediate level wastes (ILW)*

These wastes can also be extremely radioactive but they are more stable. They do not require that the heat generated by radioactive decay is taken into account as this is small. Nevertheless, they can include highly radioactive and extremely long-lived radionuclides, just as HLW does. Some of these ILWs require environmental isolation for geological time scales. Hence the concern with such factors as: possible effects of the next ice age, depth of freezing, mountain building processes, excavation by subsequent civilizations, etc.

There are a number of streams of ILW of broadly similar activity. BNFL (1985) describes them as follows:

  (i) fuel element claddings and related debris which remain after dissolution in acid or were removed prior to reprocessing;
 (ii) various sludges and ion-exchange resins from storage pond water treatment and concentrates of liquid waste streams and plant wash out;
(iii) miscellaneous wastes containing beta or gamma-active materials chiefly formed by unserviceable equipment and other items arising from operations and maintenance;
 (iv) plutonium contaminated materials (PCM); and
  (v) graphite sleeves and stainless steel components of AGR fuel assemblies.

In addition, there will be a considerable amount of decommissioning wastes and various fairly bulky objects (such as sections of pressure vessel, lumps of nuclear submarine reactor plant etc). It is envisaged that different types of conditioning and packaging are needed prior to disposal. There are also different possible disposal routes for the various types of ILWs.

*Low level wastes (LLW)*

These wastes contain sufficient radioactivity to preclude their disposal as ordinary non-radioactive refuse. The levels of activity are not supposed to exceed 4 GBq/tonne of alpha or 12 GBq/tonne for beta and gamma activity. Considerable quantities of LLWs are discharged in both liquid (via the infamous pipelines into the Irish Sea) and gaseous forms. These discharges are subject to authorizations.

Typical LLWs are low in radioactivity and high in bulk. They range from rubbish (such as gloves, packages, shoes) to some plutonium-contaminated materials with low levels of contamination. Virtually everything that is used or is in contact with radioactive substances eventually ends up as LLW; for example, JCBs used in site construction have even been buried as LLW. The volumes of LLW depend on the degree of compaction and pre-processing that is performed; this reduces bulk but also increases levels of radioactivity per unit of volume. Yet very little of this substance is considered threatening no matter how neurotic the observer may be. On the other hand, some of the wastes will remain radioactive for fairly lengthy periods of time; say 300 years or more. Land use restrictions and monitoring may well be required for a few hundred years after a LLW dump has ceased its operational life and there is always the risk that the real problems of environmental pollution will only emerge later; for example, from leaching of radionuclides into ground water acquifers.

## Storage versus disposal

A key question in dealing with radwaste is whether it is better to dispose or store the radwastes indefinitely. It would seem to be technically feasible simply to store the wastes indefinitely in above ground repositories. Indefinite storage would allow monitoring and thus avoid a potential disasterous situation in which a radwaste dump might have to be exhumed by subsequent generations to clean it up. On the other hand, given the volumes of waste being suggested in Table 2.1 and elsewhere, it is possible that indefinite surface storage might present feasibility problems and might eventually limit future nuclear activities; indeed, this may be a major reason for its support by the anti-nuclear groups. Old power reactors would have to be left on site, for ever, as radioactive monuments to the fission age. The tremendous heap of reprocessing wastes would require extensive constructional activity and new sites would have to be found. But would people be any happier living next to a Greenpeace nuclear waste warehouse than to a radwaste dump? However, if this option was considered the only publicly acceptable solution, then there might well be advantages of cost and speed of availability that would make it quite appealing at least in the short term. This approach does imply long term institutional stability and support and freedom from external events that might threaten waste containment; for example, a fire might be fairly catastrophic as might a major civil disturbance or war. Additionally, maintenance might well increase workforce radiological risks especially if the packaging started to corrode. This particular 'fear' is rather weak because if the packaging is that bad then clearly it should not be in use. However, there are arguments that it would be better to remove the stuff from the immediate human environment but whether this is best done as soon as possible or on a more relaxed timescale is certainly a point for further debate. Certainly, if delaying the disposal will significantly improve public acceptability then it deserves re-investigation.

The storage issue is complicated by a broad spectrum of differing interpretations of the distinction between storage and disposal. In essence, storage implies retrievability, continuation of maintenance and surveillance. However, all the intended waste dumps allow for a period when the waste could, theoretically, be retrieved before it becomes irreversably disposed. This is a largely semantic debate and it may be assumed that once a facility is in operation, retrievability is likely to have a very limited applicability. For instance, are the wastes in Drigg or the Atlantic retrievable right now? The answer is almost certainly yes it might be possible, but no it would be best not done. On the other hand, it is quite clear that for limited time periods (up to 100 years?) temporary surface storage might well be an attractive proposition especially if it avoids the problems of public acceptability associated with early disposal. It could be used to reduce the urgency by which radwaste dumps had to be found and made operational, and would allow for additional research into the safety of the intended depository design. This is in fact the option being used for HLW but it is difficult to imagine the scale of surface storage facility needed to handle all the wastes likely to arise by even 2030AD. For these reasons, the government clearly prefers final disposal of all wastes on the basis that indefinite storage presents unacceptable risks given that it firmly believes that safe disposal options do exist (DOE, 1986a). The key political questions that might one day be addressed are how early need early be and whether a 50- or 100-year storage period would still count as early disposal.

## Radioactive wastes in Britain

The geography of the waste generation process is of great interest. There is already a considerable movement of radioactive wastes around Britain's railway, and to some extent, road network. Table 2.3 provides estimates of the approximate magnitudes of the amount of wastes generated by a PWR station both during its operational life and afterwards. Note the large amounts of decommissioning wastes that are likely to be created. Indeed the large volumes will present major problems at some future date. Disposal will require transport and all the potential difficulties associated with it.

*Table 2.3* Waste generation by a PWR station and decommissioning wastes for MAGNOX and AGR

| Type of waste | Operational life Life of PWR | Decommissioning wastes MAGNOX | AGR | PWR |
|---|---|---|---|---|
| LLW | 57,300 | 20,000 | 12,288 | 7,000 |
| ILW | 2,100 | 8,000 | 8,598 | 1,600 |
| HLW | 100 | 0? | 0? | 0? |

Source: CEGB quoted by Layfield (1987) para. c41, p.13; para. 43.1; and Environment Committee (1986) p.xxii

Table 2.4 shows estimates of the wastes likely to be arising at each of the nuclear power sites from currently operational facilities. For any site, the real amounts will vary depending on factors that cannot be accurately predicted. The dominance of Sellafield and Dounreay is very noticeable. It should also be remembered that Sellafield is a major LLW generator but because of its Drigg dump the figures do not accurately reflect this aspect. The amount of HLW depends on whether there is reprocessing and what category spent fuel is allocated. Currently, the CEGB do not consider spent fuel to be HLW because it is capable of being reprocessed and as such it is considered to be valuable. It also means that no HLW is shown for individual power sites.

*Table 2.4* The geography of waste generation: committed raw waste arisings up to 2030 AD

| | Waste Stream: (raw waste volumes) cu.m | | | | |
| | Operational | | Decommissioning | | |
| Site: | LLW | ILW | LLW | ILW | HLW* |
|---|---|---|---|---|---|
| Amersham International (Amersham) | 57,100 | 579 | 0 | 0 | |
| Amersham International (Cardiff) | 20,100 | 678 | 0 | 0 | |
| BNFL Chapelcross | 2,250 | 74 | 21,000 | 0 | |
| BNFL Capenhurst | 732 | 55 | 250 | 0 | |
| BNFL Sellafield (including THORP) | 575,100 | 61,755 | 52,616 | 47,000 | 3,820 |
| BNFL Springfields | 195,300 | 344 | 344 | 11,400 | |
| CEGB Berkeley | 4,360 | 2,580 | 30,637 | 8,350 | |
| CEGB Bradwell | 1,800 | 898 | 30,637 | 8,350 | |
| CEGB Dungeness A and B | 6,270 | 1,913 | 51,657 | 14,350 | |
| CEGB Hartlepool | 1,960 | 1,050 | 21,020 | 6,000 | |
| CEGB Heysham 1 and 2 | 9,310 | 2,200 | 42,040 | 12,000 | |
| CEGB Hinckley Point A and B | 5,590 | 1,033 | 51,657 | 14,350 | |
| CEGB Oldbury | 5,470 | 612 | 30,637 | 8,350 | |
| CEGB Sizewell | 2,040 | 1,180 | 30,637 | 8,350 | |
| CEGB Trawsfynydd | 2,430 | 2,250 | 30,637 | 8,350 | |
| CEGB Wylfa | 2,150 | 883 | 30,637 | 8,350 | |
| SSEB Hunterston A and B | 7,470 | 4,840 | 86,300 | 26,165 | |
| SSEB Torness | 4,490 | 2,220 | 50,820 | 13,005 | |
| UKAEA Culham | 0 | 0 | 36,276 | 2,084 | |
| UKAEA Dounreay | 8,980 | 2,370 | 0 | 0 | |
| UKAEA Harwell | 44,100 | 4,670 | 0 | 0 | |
| UKAEA Risley | 284 | 0 | 0 | 0 | |
| UKAEA Winfrith | 22,500 | 1,680 | 55 | 1,540 | |

* all spent fuel from power stations is transported to Sellafield for storage and reprocessing giving rise to a single source for HLW
Source: DOE (1988) 1987 UK Radioactive Waste Inventory

## Historical highlights

The next two chapters are concerned with the development of policy and regulation (chapter 3) and some detail of the previous attempts to handle the waste disposal problem (chapter 4). This is followed (in chapter 5) by a closer look at the sites either being used as dumps or which have been suggested as being potentially suitable. The criteria used to define suitable radioactive waste sites are discussed in chapters 6 and 7. In all these chapters there is an inevitable historical dimension. The policies, regulations and siting criteria have all evolved through time. However, the large number of agencies, departments, and organizations that are concerned with radwaste makes the story a particularly confusing one. Before plunging into detail, it is useful to step back

*Table 2.5* Key historical developments related to radioactive waste

| Date | Event |
| --- | --- |
| 1952 | Chemical separation starts at Windscale |
| 1957 | Windscale fire |
| 1958 | Reprocessing plant at Dounreay |
| 1964 | New reprocessing plant at Windscale (second separation plant) |
| 1969 | Windscale Head End oxide fuel reprocessing plant (closed 1973) |
| 1971 | British Nuclear Fuels plc (BNFL) formed from part of UKAEA |
| 1977 | Windscale Public Inquiry |
| 1978 | Prospecting for HLW sites (abandoned in 1981) |
| 1982 | Formation of the Nuclear Industry Radioactive Waste Executive (NIREX) |
| 1983 | NIREX announces interest in Billingham and Elstow |
| 1983–5 | Sizewell B Public Inquiry |
| 1985 | Billingham site withdrawn on the recommendation of government |
| 1986 | Four new sites announced: Elstow, Fulbeck, Bradwell, South Killingholme |
| 1987 | All four sites abandoned |
| | 'Way Forward' public participation exercise |
| 1988 | Interest in Sellafield, Altnabraec and Dounreay |
| 1989 | Selected sites announced: Dounreay, Sellafield |
| 1992? | Windscale THORP starts 10-year life |
| 1992? | Public Inquiry into proposed deep-disposal site |
| 2005? | National radwaste dump opens |

and have a look at the total picture of events as a means of orientation. The aim here is not to provide a detailed picture of the nuclear industry but a summary of those parts that are relevant to understanding the nuclear waste story. The narrative is not meant to be purely a descriptive account of bland facts, rather it is meant to provide a critical commentary and analysis of events and policies.

*Table 2.6* Acts Of Parliament

| Date | Event |
| --- | --- |
| 1946 | The Atomic Energy Act |
| 1948 | The Radioactive Substances Act |
| 1954 | The Atomic Energy Authority Act |
| 1959 | Nuclear Installations (Licensing and Insurance) Act |
| 1960 | The Radioactive Substances Act |
| 1965 | The Nuclear Installations Act |
| 1970 | Radiological Protection Act |
| 1974 | Health and Safety at Work Act |
| | Dumping at Sea Act |
| | Control of Pollution Act |

*Table 2.7* Key policy statements

| Date | Event |
|------|-------|
| 1955 | White Paper (CMND 9389) announcing civilian nuclear power programme |
| 1959 | White Paper (CMND 884) Control of Radioactive Wastes |
| 1971 | British Nuclear Fuels plc (BNFL) formed from part of UKAEA |
| 1976 | Royal Commission on Environmental Pollution, Sixth Report: Nuclear Power and the Environment (Flowers Report) |
| 1977 | Government's Response to Flowers Report (Cmnd 6820) setting out basic principles |
| 1977 | Windscale Public Inquiry |
| 1978 | Radioactive Waste Management Committee (RWMAC) created |
| 1982 | White Paper (Cmnd 8609) said HLW not an immediate problem NIREX created |
| 1984 | Department of Environment, Radioactive Waste Management: the National Strategy |
| 1983-85 | Sizewell B Public Inquiry |
| 1986 | House of Commons Environment Committee: Radioactive Waste White Paper (CMND 9852): Government's response to Environment Committee |
| 1986 | Special Development Order, July 7th, to allow investigations to proceed. Near surface facility only for LLW |
| 1987 | Way Forward: NIREX discussion document |
| 1989 | Short list of sites announced |
| 1992? | Public Inquiry |

There is praise when praise is deserved and criticism when that is appropriate. There is some factual description but there is also an attempt to look behind the facts to obtain 'a plain-English guide' to what is a highly complex process. Perhaps this is something that only geographers (or others) not so deeply immersed in nuclear science and technology, can attempt. Whether we are successful in communicating the problems and possible avenues of solution without losing sight of the 'facts' as viewed from the nuclear industry's more narrow perspective, is something for the reader to decide. We emphasize that neither a pro- nor an anti-view is intended, rather it is meant as a dispassionate and independent analysis as 'we see it'. We make no apology if our version differs from other people's. Our perceptions are of course biased by the published material at our disposal and cannot therefore be considered to be always, necessarily correct; indeed, we make this caveat because we know what the response of those we criticize will be. Nevertheless, it is designed to inform, to provoke debate and later to act as a shop window for the use of GIS technology. Above all it is intended to be a contribution towards solving a problem which whether we like it or not is with us and has to be faced sooner rather than later.

The brief historical overview presented here is broken down into three sections: selected key nuclear developments, Acts of Parliament which establish a regulatory framework for the nuclear industry and government policy announcements. In reality they all interact and are interwoven; they are separated in the hope that thereby the information might be more readily assimilated. Table 2.5 provides a brief historical record of key nuclear developments related to radioactive waste. The story lying behind these headlines is one of initial haste, botched public relations and repeated failure to secure the objective of having a national radwaste dump in operation by 1990. It will be argued that failure was only to be expected because of the complexity of the problem. What seemed to be a fairly straightforward technical process turned out to be massively complex because of a failure to appreciate the importance of securing public acceptability and an intrinsic susceptibility to changes in policy. Table 2.6 lists the principal Acts of Parliament. These provide little of any directly relevant substance to the radioactive waste problem but they do provide a regulatory framework within which all the developments have to operate. Table 2.7 lists the principal dates of important policy documents. They relate to reviews of the nuclear industry that focus on the radioactive waste problem and also key policy statements that have had some impact on the problem and its proposed solution.

## Conclusions

Clearly, the radwaste dump story is not likely to end soon, if ever. Fairly soon new nuclear plant will be built. More nuclear reactors will be closed down and the older bits of Sellafield removed and more nuclear submarines will become available for decommissioning. Will it be much of a surprise if plans soon surface concerning near-surface facilities for large items of decommissioning waste? Will anyone be surprised when Elstow, Killingholme, Billingham, Bradwell and Fulbeck are again connected with radwaste dump interest? If it does not happen soon, there is every prospect of it happening before too long. Do we really know what we are letting ourselves in for? All this fuss, so soon in the nuclear electricity age, when less than 20 per cent of the generation is nuclear, does not augur well for the 21st century when 50, maybe even 80 per cent will become nuclear. Does it matter, all this fear and concern about something that while rare today can only be expected to become commonplace and widely accepted within 100 years? Can we stop the nuclear era merely because people do not like it and fear what they do not understand? Maybe what we see today are the last remnants of anti-nuclear opposition before dependency becomes so great that continued opposition becomes treasonable! Maybe it is a feeling of being trapped by an out of control, rudderless machine racing forward at an ever increasing speed, disregarding everything in its path. Should we join it, or oppose it? Maybe too many people are worrying about the wrong things, they are too sensitive and need to adopt a more relaxed, neutral view. Perhaps, more people need to come to terms with the realities of the nuclear world in which we live. It matters not whether there are any British reactors when incredibly bad accidents anywhere in Europe could be almost as bad; and political parties in Britain will never be able to have much or any control over those! So perhaps we need to learn to love what many have started to hate. Why? Because we have little or no choice in the matter. The only real prospect for the future is to insist that the nuclear age is perfected, any faults in the technology eliminated, and the best and safest possible 'King' of radwaste dumps developed. If the technology cannot be avoided, then let us make sure that it is as safe as possible. Siting is therefore a most important activity.

# 3

# The policy story: rules, regulations, and White Papers

This chapter takes a broad look at policy matters. It attempts to unscramble the complexity and identify the bare essentials. The subject is rather detailed and in places boring, but it is important as a means of understanding the situation that exists today. It is also significant because policy matters provide the principal interface between politicians and the nuclear industry. It is relevant then to set the scene with a particularly critical quote from the Environment Committee's (1986) review of the Radioactive Waste problem.

'It had become apparent to us that far from there being well-defined, publicly debated policy on the creation, management and disposal of radioactive waste, there was confusion, and obfuscation among the various organisations entrusted with its care.' (p.xii). They continue: 'In short, the UK government and nuclear industry are confused. On the one hand, bold announcements about prospective new disposal sites are issued. They are then withdrawn, left hanging in the air, or modified *ad hoc*. On the other hand, a very large proportion of radioactive waste goes on being produced unquestioned and a sequence of different studies show that the UK is still only feeling its way towards a coherent policy. For an issue of such great public concern, this is regrettably inadequate.' [p.xii].

Strong words, it is a pity that the Committee were unable to justify their concern fully. Nevertheless, in the process they did succeed in opening up this area to public view and political debate and many of the less contentious of their 43 recommendations were accepted. Sadly, they were beaten by the invisible pro-nuclear lobby, but not without a fight. The interested reader is referred to the three volumes that make up the *First Report of the House of Commons Environment Committee on Radioactive Waste* (HMSO, 1986). The government's reply is also essential reading (Cmnd 9852). This chapter tries to justify the Environment Committee's concern.

How did the situation described by the Environment Committee come about? What was the nature of the regulatory and policy framework that would consider it quite normal and which also affects the organisational and

39

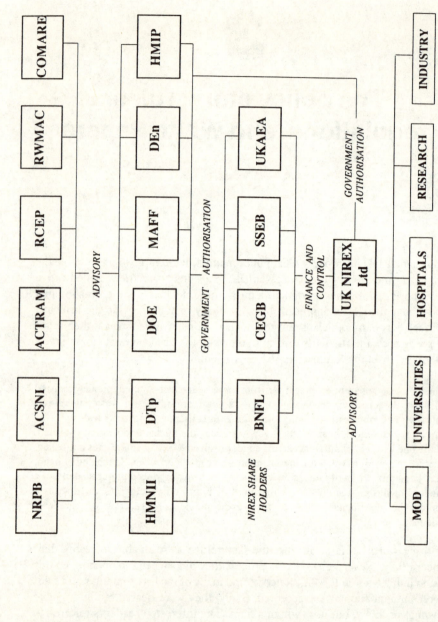

*Figure 3.1* Organization chart of waste management agencies and advisory committees in 1988

institutional structures relevant to locating a radwaste dump? The complexity arises from the fact that radioactive waste management in the United Kingdom is shared between government, regulatory and advisory agencies, and the nuclear industry. Government is responsible for developing and implementing a long term national strategy for radioactive waste management, while the nuclear industry is expected to provide the infrastructure and pay all the costs. This activity takes place within a setting which is still evolving; the latest part of which will only start when the effects of privatizing electricity begin to influence the nuclear industry's radwaste executive. Figure 3.1 provides an organizational chart of waste management agencies and advisory committees in 1988. There are 31 different agencies listed. Where to begin is not easy. Here attention is focused first on the Acts of Parliament that establish a regulatory framework for the control and management of both the nuclear industry and nuclear waste. This is followed by a more detailed look at the emergence of policy.

## What does the law say?

A series of Acts of Parliament control the creation, storage, use and discharge of all radioactive substances. It is useful to examine what light, if any, they throw on the radioactive waste disposal problem. It is particularly interesting to look for any signs to indicate that the radwaste disposal problem was foreseen and is, therefore, explicitly rather than implicitly embodied in the various Acts of Parliament. Atomic energy has always been subject to legislation, usually far more stringent than that which applies to non-nuclear activities that also involve hazardous substances. The legislation has been primarily concerned with three aspects: (1) granting the relevant minister powers to do virtually anthing considered appropriate while also ensuring non-disclosure of any details considered of importance for defence; (2) to establish a system of licences and controls relating to the storage and use of radioactive substances; and (3) to establish financial compensation and ensure insurance related to litigation arising from any breach of duty.

The Atomic Energy Act (1946) begins this process by granting the Minister of Supply wide ranging powers 'to produce, use and dispose of atomic energy and carry out research into any matters connected therewith' (Section 2(a)); to manufacture, buy, or acquire, store and transport related articles; and to 'do all such things (including the erection of buildings and the execution of works and the working of minerals) as appear to the Minister necessary or expedient for the exercise of these foregoing powers' (Section 2(c)). To this end there are provisions for compensation relating to compulsory land purchase. There is no explicit mention of waste.

The Radioactive Substances Act (1948) granted the Minister of Supply power 'to manufacture or otherwise produce, buy or otherwise acquire, treat, store, transport and dispose of any radioactive substances; and to do all such things (including the erection of buildings and the execution of works) as appear to the Minister necessary or expedient.' (Section 1). This Act also established a system of licences to control the sale, supply , and transport of radioactive substances. There is no specific mention of waste. It was during the period 1948–52 that the Windscale Piles were constructed, the first Windscale reprocessing plant built, and the first Windscale waste pipe into the Irish Sea became operational. Military urgency seems to have ensured minimal details being published about any nasty problems relating to radioactive pollution. It is also probable that the scientists themselves did not fully understand the processes involved or the potential nature of the long term health threat. At the time, the Minister of Supply could (in the national interest) do more or less anything considered expedient to further the development of atomic weapons. There were compulsory purchase powers and intense security. It is interesting that when the Nuclear Industry Radioactive Waste Executive (NIREX) was established in 1982 these compulsory purchase powers were not extended to it. If the Minister of Supply (and the Atomic Energy Authority) had been given the task of radioactive waste disposal then it could simply have (1) used national security arguments to hush everything up, (2) used compulsory purchasing powers to acquire whatever sites were necessary, and (3) seen no need for any planning inquiries or public participation exercises. In Britain, the Official Secrets Acts still effectively conceal whatever waste disposal facilities and practices were developed in the early atomic era. No amount of official denials will ever constitute proof that all the long-lived military wastes have in fact been stored and are included in the current inventories; indeed the precise nature of the military contribution is still a secret although it is claimed by the Ministry of Defence (MOD) to amount to less than 20 per cent of the civilian stocks. It can only be speculated (there is no evidence) that, maybe, much of these early wastes were in fact merely diluted and dispersed through the pipeline(s) into the Irish Sea or dumped in the Atlantic without any apparent ill-effect. Knowledge of this secret activity, assuming of course it did occur, may help explain the continued and persistent desire of the British government to re-commence dumping in the Atlantic (see Chapter 4). Whatever the truth, it is no more than a minor historical diversion since the main theme of this book is what to do with current and future wastes; assume that sea-dumping is not a legitimate option even if it seems acceptable from a scientific point of view.

The Atomic Energy Authority Act (1954) provided for setting up an Atomic Energy Authority with the functions of developing and carrying out research into atomic energy and of producing radioactive substances. The functions of the authority are limited to the research part of the 1946 Act and not explicitly to atomic weapons production, the latter functions remained the responsibility of the Minister of Supply and continued therefore to be exempt from legislation. One interesting subsection gives the Atomic Energy Authority rights 'to place any pipe across land, whether above or below ground, and to use, repair, and maintain that pipe, without purchasing any other interest in the land' (Section 5(1)). No doubt an oblique reference to the ongoing contemporary interest in piping radioactive wastes out to sea! This Act also established that it was the duty of the Atomic Energy Authority to ensure that no ionizing radiations from their premises or waste discharges cause any hurt to any person or any damage to any property. Additionally, radioactive waste discharges had to be authorized. Still no explicit mention of radioactive wastes except for discharge authorizations. Clearly, radioactive wastes were still being treated as being similar to sewerage and that disposal was little more than just another authorized release that would eventually be diluted and dispersed.

Shortly after this Act received royal assent, the White Paper advertising the creation of a civilian nuclear power programme was published. The diffusion of atomic energy into the public and commercial sector was to result in a belated flurry of legislation designed to control the process; belated because the construction programme preceeded the Acts of Parliament that established a regulatory framework for it. The Nuclear Installations (Licensing and Insurance) Act (1959) created a system of licences to control in the interests of safety the building and operation of nuclear reactors and related installations. A nuclear licence was necessary before any site could be used for the production or use of atomic energy, included was the storage, processing and disposal of nuclear fuel and bulk quantities of radioactive materials (see Section 1(b)). It also imposed on any person granted a licence for the use of a site for building or operating a nuclear installation a similar absolute liability to Section 5(3) of the Atomic Energy Authority Act (1954) for all hurt to persons or damage to property caused by radiation from the site even if due to unavoidable accidents, together with an obligation to take out insurance cover for up to £5 million in respect of any one occurrence on any one site. By this, radioactive waste disposal facilities (if they had existed) would have been treated in the same fashion as any other nuclear installation.

The Radioactive Substances Act (1960) further regulated the keeping and use of radioactive material, including the accumulation and disposal of radioactive waste – the first time there is any explicit mention of this aspect of the atomic era. This Act made it unlawful to dispose of or to accumulate radioactive wastes except in accordance with authorizations given by the appropriate minister or ministers. But authorizations are not required for the accumulation of wastes on premises of the Atomic Energy Authority or for any sites with a nuclear site licence. Moreover, the minister may by order exempt particular descriptions of waste from the authorization process. All very necessary but somewhat vague. There are no details of how wastes may be disposed. There is seemingly no concept of a waste dump. Instead generic responsibilities are established: '... no person shall, except in accordance with an authorisation granted in that behalf under this subsection, dispose of any radioactive waste on or from any premises which are used for the purposes of an undertaking carried on by him' (Section 6(i)). However, this 1960 Act also states that, 'If it appears to the Minister that adequate facilities are not available for the safe disposal or accumulation of radioactive waste, the Minister may provide such facilities, or may arrange for the provision thereof by such persons as the Minister may think fit' (Section 10(1)). There is no suggestion that this task might be difficult, although there is a requirement to consult with local authorities and to make a reasonable charge for use of the facilities. The minister may also dispose of waste where for any reason it is unlikely to be lawfully disposed of. This section provides a legal basis for the subsequent creation of NIREX and gives responsibility for research and monitoring to the relevant government department.

The Nuclear Installations Act (1965) clarifies some of the duties associated with nuclear site licences and replaced the 1959 Act. There is no specific mention of radioactive waste although a radioactive waste facility would be covered by this Act. Indeed, it is currently intended that the control of any radwaste dump would be handled by authorizations under the 1960 Act with licences granted under the 1965 Act. There is an interesting potential anomaly here in that the 1965 Act would continue to apply to a radwaste dump for ever; or until 'there has ceased to be any danger from ionising radiations from anything on the site' (Section 5(3)). This may presumably suggest that a radioactive waste dump would constitute a perpetual liability for the licensee, unless of course the dump was considered so safe as to no longer constitute a danger once its operational life had finished. However, in principle at least this responsibility could become of infinite duration. The long life of nuclear waste facilities measured in geological timescales would seemingly require the perpetual application of the relevant parliamentary statutes although this appears to be an issue not explicitly dealt with by law; no doubt (in due course) some amendment will be required to limit the period of application. At present it appears that a nuclear waste disposal site would need a nuclear site licence for ever and no one seems to have considered all the implications of that. However, the principal ongoing responsibilities are not onerous and relate to: ensuring that no ionizing radiations emitted from the site cause any harm or damage to property, a liability for compensation, special insurance cover provisions, and fencing (and any other provisions suggested by the minister) to prevent or warn of any risk of injury from the site. The minister can however set at any time conditions to the nuclear site licence regarding the handling and disposal of nuclear material, and this might be viewed as a problem of undefined extent. Another potential anomaly might occur if the wastes were to be disposed in such a way that they could still be considered to be in storage. This would then only require permission under the 1960 Act since no environmental discharge was being involved.

So much for the law, which seems to exempt from control most of the activities related to the development of atomic weapons and to be so vaguely worded that it fails to mention anything remotely resembling a radwaste dump. On the other hand, these very general laws seem to provide such blanket coverage that they give the relevant government departments total power to do virtually anything they consider appropriate. In this way, the key responsibilities rest with the authorizing departments and the advice they receive from various related agencies. Whatever they decide has the force of the law behind it. It is difficult to imagine other areas of legislation where government departments have been given so much power. However, it can no doubt be justified here because of the unique nature of the nuclear industry. It would work very well if the principal bodies were able to operate in a truly independent fashion; in practice, the pro-nuclear lobby is able to influence most of those involved. It obviously helps having a large number of 'independent' experts who were once employed in or by the nuclear business and who still believe in the glory of the all nuclear dream. This does not stop all opposition or prohibit heretical statements; but it does tend to ensure that the key decisions are nearly always nuclear-favourable. Of course, there is no proof of any of this, other than the occasionally inexplicable behaviour of various key independent advisory bodies.

So nuclear waste facilities are regarded in the same manner as nuclear power stations and this is probably the best that can be expected. Given the very limited life of reactors compared with waste depositories, the statutory framework is fairly stringent albeit lacking in details as to the precise safeguards to be applied. The only real omission relates to the need for detailed records and inventories to be kept of the precise nature of the materials being stored or disposed of. These would be of incredible value at some future date when, as is virtually inevitable, a subsequent society decides to exhume a radwaste dump in order to make it safe using vastly improved technology. It is one thing leaving a radwaste dump for future generations to sort out, it is quite another to fail to document properly the problem that we leave behind.

## The evolution of radioactive waste regulations

The controversy over radioactive waste is relatively recent in origin and this is reflected in the nuclear legislation that has just been reviewed. The early decades of nuclear power received general public and all party parliamentary acceptability mainly because of the perceived need for nuclear weapons and the popular view that here was a new technology which promised a bright new future of clean, plentiful and cheap electricity. A technology which would guarantee unlimited power and enhance Britain as a super power was given 'jet age' status and a degree of charisma that allowed the nuclear industry to get on with the necessary developments. Not suprisingly, nuclear waste was not a major priority for the young, vibrant, rapidly expanding and self-confident civilian nuclear industry. The waste issue was much less glamorous than the prestigious projects associated with the earlier stages of the fuel cycle and the industry believed that no technical problems would be encountered in the ultimate disposal of radioactive wastes when it became necessary to do so (BNFL, 1985). Meanwhile, most of the wastes were being dumped either at Drigg (LLW) or in the north east Atlantic (LLW & ILW). That is not to say that the ultimate disposal of waste was neglected. Appropriate legislation was enacted, primarily in the wake of the first policy guidelines written for the control of nuclear waste in the White Paper of 1959 (Cmnd 884), and this had resulted in the 1960 Radioactive Substances Act. Indeed, both major Acts from this period – the Radioactive Substances Act, 1960 and the Nuclear Installations Act, 1965 are still the pillars of nuclear regulation today.

The internationalization of nuclear power also led to the establishment of international agencies and conventions to advise and provide guidelines for the membership. In the 1950s and early 1960s the United Nations (UN) established UNSCEAR (UN Scientific Committee on the Effects of Atomic Research), the International Atomic Energy Agency (IAEA) was created and EURATOM became the nuclear agency of the newly formed EEC. Nevertheless, it was not until the 1970s that these agencies began to establish radioactive waste management divisions and began to devote more resources to this problem than to nuclear safety and radiological protection which always seemed to be of more immediate importance (Blowers and Lowry, 1985). This approach was mirrored in the United Kingdom where the only new piece of legislation referring to radioactive wastes was the Radiological Protection Act of 1970 which created the National Radiological Protection Board (NRPB), an advisory body on establishing radiobiological standards. Nevertheless, the NRPB, like most of its international counterparts, used the International Commission on Radiological Protection (ICRP) standards as the basis of its work in the United Kingdom. This commission, founded in 1928, is the recognized setter of standards for radiological protection although it has come in for criticism in the last decade, especially in the United States (see Layfield 1987, chapter 30 for an excellent review).

By the 1970s growing public concern at the less attractive features of nuclear power either followed or gave rise to a populist environmental movement which began to have ramifications for the nuclear industry throughout the world. The accident at Three Mile Island in the late 1970s suddenly and very dramatically drew attention to the full horror potential of a nuclear technology that appeared to be increasingly out of control and a law unto itself. Initially, environmental groups marshalled their efforts on aspects of nuclear safety; for example, the risks of reactor accidents and the possible health effects of low level radiation. However, by the late 1970s attention was being focused on radioactive waste management. In 1976 the famous Sixth Report of the Royal Commission on Environmental Pollution called *Nuclear Power and the Environment* (the Flowers Report) and the following year the Windscale Public Inquiry undoubtedly helped draw attention to the 'dirty' back-end part of the nuclear fuel cycle. This public interest also coincided with the creation of the Radioactive Waste Management divisions within international agencies. International conventions started to appear as a means of controlling or eliminating 'commons-type' pollution problems. The best example was the decision to suspend and then end dumping radioactive wastes at sea in 1983.

The 1970s in the Uunited Kingdom saw more legislation which impinged upon nuclear waste management, namely the Control of Pollution Act, 1974, the Health and Safety at Work Act, 1974, and the Dumping at Sea Act, 1974. However, the basic regulations laid down in 1960 and 1965 still apply today although the divisions of responsibility have been redrawn over time. The lead agency which has prime responsibility for overall policy and strategy is now the Department of the Environment (DOE) and it shares responsibility with the Ministry of Agriculture, Food and Fisheries (MAFF) for authorizing disposal or discharges of radioactive waste. In essence, these government departments set conditions on the authorizations, ensure compliance, carry out R&D, assess radiological implications of disposals and approve new disposal policies and sites. The DOE through its Radiochemical Directorate regulates waste discharges on land while MAFF is responsible for monitoring radioactivity through the food chain and its brief includes regulating waste disposal at sea under the Dumping at Sea Act. Another government department, the Health and Safety Executive (HSE), is responsible for atmospheric discharges. Incidentally, in the new administration restructuring of 1974, the NII was subsumed into the HSE. Other government departments are involved in radioactive waste matters either through regulatory obligations (the Department of Transport (DT) has to ensure that operations comply with IAEA/DT guidelines in the movement of wastes) or a more general brief such as Department of Health and Social Security (DHSS) concern for the health repercussions of radioactive discharges. With time, more and more specialist advisory bodies are drawn into the organizational structure of the radwaste responsibilities of government. Another look at Figure 3.1 might be useful at this point.

## The policy background to the management of solid radioactive wastes

To understand the basis of government policy on the disposal of low and intermediate level wastes it is very useful to trace their development over the last decade or so. The nuclear waste issue did not surface as a politically contentious one until the mid-1970s, when the Sixth Report of the Royal Commission on Environmental Pollution, the Flowers Report, was published. Prior to that time, radwaste disposal was seemingly not a problem; it was either sent to Drigg or dumped in the Atlantic or stored as high level waste at Sellafield. The small size of Britain's nuclear power programme was insufficient to draw much attention to it. It was probably also convenient to keep it quiet as efforts were made to develop a much larger-scale nuclear power programme. This implicit do-nothing 'strategy' would have continued working were it not for the massive scale of the planned expansion of nuclear power and particularly the eagerly anticipated development of fast breeder reactors with the consequential need for a much larger scale of reprocessing and waste disposal than was currently being performed out of the public and media gaze. The subject was still largely restricted and the current 'open-door' policy was not yet in place. Some of the popular newspapers began to run leaders on how Britain was about to become the nuclear dustbin of the world; a reference to contracts being negotiated between BNFL and other nations to reprocess their oxide fuel. This was further stimulated by the Windscale Public Inquiry in 1977 concerning planning permission to construct the Thermal Oxide Reprocessing Plant (THORP) at Sellafield.

*Flowers Report*

The Flowers Report is usually heralded as the start of the radwaste dump era. It argued that waste management objectives should be clearly identified at the outset of a nuclear programme rather than at a later stage where important options could be foreclosed. The problem had been recognized from the very onset of the nuclear power programme but it was not until 1976 and the Flowers Report that the structural beginnings of a national strategy for the safe disposal of radioactive wastes began to emerge. The report pointed out that neither government departments nor other organizations had sufficiently appreciated the need to make adequate arrangements for radioactive waste disposal in good time before the need arose. The report drew attention to the need to develop environmentally acceptable means of disposing of the increasing quantities of wastes being generated and concluded that: 'there should be no commitment to a large programme of nuclear fission until it has been demonstrated beyond reasonable doubt that a method exists to ensure the safe containment of long-lived highly radioactive wastes for the indefinite future.' (para. 533)

The main cause of Flowers' concern was the then United Kingdom Atomic Energy Authority's view that 104 GWe of nuclear capacity would be required by 2000 (equivalent to around 100 PWRs!). It is now likely that the overall nuclear capacity will be only 10 to 15 per cent of that estimate. Nevertheless, the Flowers Report emphasized the need for research to establish the Best Practicable Environmental Option (BPEO) for dealing with each waste stream. The Royal Commission also recommended that two new administrative bodies should be created to foster a more coherent waste management strategy: an advisory body which would monitor waste management practices and report to government (the Nuclear Waste Management Advisory Committee) and an executive independent statutory body to oversee the whole process of waste disposal (Nuclear Waste Disposal Corporation). Both bodies were to call on a range of expertise, including a strong environmental representation. In 1978 the Radioactive Waste Management Advisory Committee (RWMAC) was established to provide independent (of the nuclear industry) advice on the development and execution of policy for the management of civilian radioactive waste but without any statutory or executive functions. However, the government at that time did not reach a decision on the recommendation that a Nuclear Waste Disposal Corporation should be created. It is also ironic that Flowers drew attention to HLW but not the LLW and ILW that is now causing such concern.

*1982 White Paper*

In response to the Flowers Report, the government produced its policy objectives in 1977 (Cmnd 6820) which were further endorsed by a new government in 1982 (Cmnd 8607). These objectives were:

(1) to minimize the creation of waste from nuclear activity;
(2) to deal with waste management problems in principle before any large scale programme is undertaken;
(3) to carry out the handling and treatment of wastes with due regard to environmental considerations;
(4) to dispose of wastes at nuclear sites in accordance with a programme;
(5) to provide adequate research and development on methods of disposal; and
(6) to dispose of wastes in appropriate ways, at appropriate times and in appropriate places.

It is noted the responsibility for safeguarding the environment was given to a different government department (Department of Environment) from that responsible for promoting nuclear power (Department of Energy). This was supposed to avoid a conflation of interests when the principal nuclear energy sponsoring department (the Department of Energy) was also the principal nuclear regulatory agency. In practice, it probably made little difference since both departments would have call on the same body of 'independent' experts. However it looks better to outsiders!

The six objectives are sufficiently vague that they can be interpreted in fundamentally different ways. For example, environmental groups cannot reconcile how the government can attempt to minimize wastes (objective 1) when it supports the ordering of nuclear reactors in advance of need (for example, Sizewell B) and continues to accept reprocessing (the main generator of waste) as an integral part of its nuclear strategy. On the other hand, it is also apparent that the meaning of this objective was not meant to be taken quite so literally and relates to minimization of waste by optimization of waste management procedures rather than the elimination of wastes *per se* by abandoning nuclear fuel reprocessing or even nuclear power itself. Yet there should have been a public debate as to why reprocessing was necessary in the first place. Instead there has been a *de facto fait accompli*. However, it was not a dark or sinister plot by evil minded nuclear scientists but a very sensible decision when it was made in the 1950s. As chapter 2 argues, the MAGNOX fuel has to be reprocessed because it could not be stored under water for long periods of time. The same applies to the AGR fuel currently stored at Sellafield pending THORP reprocessing in the 1990s. The development of dry spent fuel stores has become possible only latterly. Thus key environmental questions are also perceived by government more as matters of nuclear economics for the nuclear industry itself rather than of major environmental significance. This neglect of the environmental implications must really be regarded as a mistake.

On the question of improved administrative procedures, the government did make the Secretary of State for the Environment (together with the Secretaries of State for Scotland and Wales) responsible for civil radioactive waste management (1977 White Paper, Cmnd 6820), while the Secretary of State for Defence is responsible for the management of wastes from the defence programme. The 1982 White Paper notes that 'the publication of information about these has to be restricted for security reasons' (p.8). Whatever arrangements are being made with respect to military wastes they are potentially: (a) secret and (b) exempt from the same constraints as apply to civilian wastes. The government also created the Radioactive Waste Management Advisory Committee (RWMAC) in 1978 and the Nuclear Industry Radioactive Waste Executive (NIREX) in 1982. But contrary to the recommendations of the Flowers Report, the membership of these bodies was drawn primarily from natural science disciplines who might be expected to both know about nuclear matters and be favourably inclined, in at least a scientific manner, towards the nuclear industry. Indeed, the nuclear industry had membership in both organizations with NIREX being no more than the waste disposal arm of its partners – BNFL, the Central Electricity Generating Board (CEGB), the South of Scotland Electricity Board (SSEB) and the UK Atomic Energy Authority (UKAEA). Indeed NIREX Ltd is a paper company with all its staff being seconded from the nuclear industry and its costs met by its sponsors. There is nothing sinister in this fact. NIREX was created, as the 1982 White Paper explains, 'to provide a mechanism by which they [the nuclear industry] can successfully fulfil their own responsibilities in this field and work within a comprehensive plan for waste management' (p.17). The government wishes therefore that the polluters should organise and pay for their own pollution control measures, subject to the rules and regulations established by government to safeguard the public. This 'polluter should pay' principle is more or less what was intended in the various Acts of Parliament regulating the nuclear industry. There was always a presumption that the industry itself should be responsible for the safe disposal of its wastes. It is not (as many seem to think) only a more recent product of Thatcherism. NIREX is considered in greater detail in chapter 4.

The 1982 White Paper is important because it confirmed the previous policy of the Labour government and mapped out a three element organization for waste management: the government, the nuclear industry and the generating boards, and the private sector. The government operates through its regulatory bodies and the 1960 and 1965 Acts. The Secretaries of State (Environment, Scotland, and Wales) are responsible for the overall strategy of waste management. The implementation of the strategy is seen as the responsibility of the nuclear industry and generating boards acting through NIREX. Finally, there is considered to be increased scope for the use of the private sector in the implementation of the strategy, particularly in design and construction.

*Drilling for HLW research abandoned*

Initially the concern expressed in the Flowers Report prompted the DOE to begin a programme of test drilling in potential host rock structures for highly active wastes. To some extent this was precisely the wrong problem to tackle first. So it was hardly catastrophic when hostile opposition from rural communities in Ayrshire, mid-Wales and Northumberland to test drilling sites caused the abandonment of these plans in 1981. However this activity generally increased public fears and paved the way for the even greater levels of public neurosis and anxiety that were to follow. Nevertheless, in 1982 the government in its White Paper (Cmnd 8607) claimed that the objective (2) of its waste management policy had been fulfilled in that the technical feasibility of disposal by burial in deep geological formations had been established in principle, a fact that was undoubtedly known before the drilling programme commenced. By now, it had been decided from available scientific evidence that these wastes could be vitrified and stored for up to 50 years prior to ultimate disposal (RWMAC, 1980, 1981). More recent work by a team of international geologists endorse this approach advocating interim storage for up to 100 years to allow for radioactive decay and a greater reduction in the heat output from the wastes (Fyfe *et al*, 1984). However, while deferral is currently popular, there is little support for the idea that indefinite storage is a substitute for disposal of HLW. The question remains as to why a similar storage-before-eventual-disposal strategy was not also applied to ILW and probably also LLW, as a device for diffusing public concern and by allowing a 50 year 'education period'. With 20-20 hindsight, it would have been the ideal solution!

Despite the abandonment of the drilling programme, the government was able to move from a situation in the mid-1970s where it had virtually no coherent radwaste policy to a situation in the early 1980s where, despite an absence of sites and disposal facilities, it was able to declare that 'Waste management is not therefore a barrier to the further development of nuclear power as now foreseen' (DOE, 1982, p.10). This confidence partly reflects a more moderate rate of expansion of nuclear power than hitherto envisaged and confidence that after five years of research, 'there is no evidence of any major scientific problems and the government has concluded that it is feasible to manage and dispose of all the wastes envisaged in the UK, in acceptable ways. There is an extensive body of existing knowledge about the technology involved. In some respects this will have to be further refined and developed, and the necessary work is in hand' (DOE, 1982 p.7). Despite the seemingly fatuous nature of this statement at a time when no facilities existed and with no sites available for development, it is nevertheless undoubtedly correct. At least in principle, there are now practicable and feasible engineering options for disposal that are already clear at least in outline. There are seemingly no great technical obstacles in general terms. However, this does not necessarily mean that there are today any solutions that meet the additional criteria of public acceptability or that they are also good enough at present to withstand a far ranging public inquiry when faced with well-informed objectors.

*Dumping suspended*

The attention of government switched away from high level wastes with the realization that the main and immediately pressing problem was for the disposal of intermediate level wastes. The first annual report of the RWMAC (1980) drew attention to 'an urgent need to proceed with identification and development of treatment and disposal options for these [low and intermediate] wastes' (p.6). Their second annual report in 1981 gave further emphasis to this aspect. Not surprisingly this view is also echoed in the 1982 White Paper which reports, 'The lack of suitable disposal facilities for intermediate-level wastes is the major current gap in waste management, and it is important that it should be remedied' (p.13). NIREX was given the role of managing this task in 1982.

The events in 1983 were soon to indicate the difficulty, almost impossibility, of this task. Issues relating to public acceptability which had been simmering away for several years were suddenly thrust into the open creating massively damaging adverse publicity for the nuclear industry. In the preceeding few years, Britain had been continuously criticized at international meetings because of its continued sea dumping programme and the levels of pollution from its Sellafield reprocessing plant. In February 1983, the London Dumping Convention resolved to prohibit sea dumping until scientific evidence could conclusively show that no harm to the marine environment would ensue – the first application of the 'guilty until proven innocent' strategy which is the reverse of normal practice. When it became clear that Britain was going to ignore this resolution, the transport unions refused to handle radioactive waste. Ultimately, the government agreed to a committee of investigation into sea dumping. The Holliday Report published a year later recommended a continued moratorium until further evidence could substantiate a case for the removal of the ban. Furthermore, it also recommended that a comparison of sea disposal with land based alternatives should be made to identify the best practicable environmental option (BPEO) for various waste streams (see chapter 5). However, it is also important to realize that sea dumping was regarded as being of limited long term utility simply because it was only thought suitable for 15 per cent of the expected 2030 arisings of ILW (DOE, 1985b, p.3).

*LLW and ILW sites announced, 1983*

By late 1983 the nuclear industry was being placed under the public microscope from a number of new angles. The Sizewell B PWR Public Inquiry was well underway and both the government and the nuclear industry were keen for all the issues to be explored in public. It was to be accompanied by a new 'open door' policy and a massive flow of hitherto restricted information became publicly available for the first time. It was during this period that NIREX announced its selection of the Billingham and Elstow sites for potential land based waste repositories for long-lived ILW, and short-lived ILW and LLW respectively. This was clearly a rushed job as the site selection predates various DOE studies that should logically have preceeded it. It was almost as if the various parties had become desynchronized. For instance, the site announcement coincided with the publication of the DOE's (1983) consultative document on the principles for the protection of the human environment in the development of such sites. This was bad enough, but the NIREX sites did not meet all the draft principles; for instance, by failing to provide a small number of alternatives for comparison purposes. There was no discussion of alternative disposal options and it was not until 1985 that best practicable environmental options (BEPO) were defined for these wastes. Despite these shortcomings, the Secretary of State for the Environment was able to claim that the effective disposal of nuclear waste, 'in ways which have been shown to be safe, is well within the scope of modern technology' (Hansard, 25 October 1983). As for intermediate level waste the government declared that 'land-disposal ... is the safest and best method, provided that a site can be found with sufficient geological certainty and stability which will remain safe for the necessary period of time' (Hansard, 25 October 1983). Presumably what was meant was that the theoretical technology could be made operational and the basic guidance principles developed retrospectively and applied *post hoc* during the actual development of the first sites; a similar procedure had been used with the first MAGNOX stations. It all looked far worse than it actually was; indeed, even if these developments had proceeded there would probably have been little risk of failure. This is, however, less than might be considered politically or publicly acceptable or a desirable basis for the establishment of a facility of such crucial importance and long-term significance.

It was within the political (and not a technical) arena that plans by NIREX for land disposal were being evaluated and this was quite different from anything the nuclear industry had previously encountered. It was hardly a surprise that the areas selected for investigation by NIREX were quick to marshall opposition to these plans and campaigns were activated to ensure all-party support. The opponents to the schemes were undoubtedly greatly encouraged by the almost arrogant manner in which NIREX handled the issues. They claimed that as many as 150 sites had been evaluated, but for commercial reasons would not name one. NIREX seemed to adopt the same site evaluation strategy that had previously characterized the CEGB's operations, probably because they used the same staff. First you identify the site(s) you want on narrowly defined engineering grounds, then you evaluate the preferred sites in detail to ensure that nothing has been overlooked. Finally, you hold a public inquiry and then invoke national interest arguments to ensure ultimate success. This strategy is virtually invincible given the British planning process but it overlooked one fundamental fact: namely, that there is no great political support whatsoever for radioactive waste dumps. No political party or MP is going to risk much for a waste dump. The fact that public opinion has for a long time now been anti-nuclear had not previously mattered. But now it did.

## A national strategy

Ironically, both the anti campaigns were called BAND (Bedfordshire/ Billingham Against Nuclear Dumping). This led to constant questioning of the Environment Secretary in the House of Commons and a full debate on nuclear waste on 3 May 1984. The government published its national strategy for radioactive waste management two months later (DOE, 1984). Unfortunately, it came nearly one year after the identification of the first shallow burial site for the disposal of short lived ILW at Elstow. This 1984 strategy document reinforced the government's commitment to developing waste repositories for LLW and ILW but delaying HLW to allow for radioactive decay. It was a very general statement and largely avoided any details. It simply stated that 'systems engineering studies by DOE identified a shallow burial and a deep burial facility for intermediate level wastes as crucial elements in a national waste management system'. For LLW and ILW, it stated that 'there is no requirement for lengthy research and development into methods of disposal', arguing that 'disposal of intermediate-level wastes has been validated as a safe option by Research and Development and by experiment or actual disposal and subsequent monitoring in several countries' (para.21). It seems likely that these assertions would be difficult to justify and the recommendations run counter to both the Flowers Report and the Royal Commission's Tenth Report (1984). It is possible that RWMAC was largely responsible. However, the 'no need for more research' argument is probably justified given the early design stage of the process, the questions of design detail could be answered later as and when they arose. It may also have reflected increasing urgency and growing panic over the loss of the sea dumping route. Moreover, it was very much a technical assessment (like most preceding policy statements) that totally ignored the realities of the political debate.

## BPEO 1986

The concept of Best Practicable Environmental Options (BPEOs) originated in the Flowers Report, and is discussed further in DOE (1986b). It implies that decisions on the storage and disposal of wastes should be based on an assessment of a wide range of occupational and environmental risks as well as the costs and social implications of the available options. BPEOs offer what may be considered as a broadly based form of cost-benefit analysis. In some ways the name is misleading because it is a decision making aid rather than a decision making technique.

The philosophy works as follows. It is assumed that the generic or site independent characteristics of various storage and disposal options can be identified and quantified. The safety assessment objective is to minimise the risks from waste disposal subject to the principle that exposures should be kept as low as reasonably achievable, social and economic factors being taken into account. The aim is to dispose of each type of waste in the most appropriate way and at the most appropriate time. So it is not an unconstrained minimization of risks which might be infinitely expensive, but a constrained optimization that seeks a balance between risk minimization and what can be justified. The DOE (1986b) document uses a multi-attribute decision analysis procedure to compare options and establish BPEOs for LLW and ILW. It should not be overlooked that the results are entirely assumption dependent, and because some or many of the assumptions are completely general and site independent, the utility of the conclusions can be questioned. This approach also assumes that these general disposal options decisions can be made prior to site specific details being available, this we would question. Indeed it is more than likely that the introduction of site specific parameters will destroy the original BPEO basis.

Nevertheless, DOE (1986b) came to a number of conclusions that are very relevant. It is stated that for every type of waste there are a number of alternative feasible options for disposal and storage. However, 'The BPEO for most LLW and some short-lived ILW is near-surface disposal, as soon as practicable, in appropriately designed trenches. ILW with more alpha-emitting radionuclides than acceptable for near-surface disposal or sea disposal will require deep underground disposal' (p.3). It continues by suggesting that, 'The amount of ILW for which near-surface disposal in an engineered trench might be preferable to deep disposal will depend on the relative importance attached to reducing costs and to reducing risks' (p.3). Since the costs and the risks are experienced by different parties, then here is the essence of a very dangerous approach. The BPEO study merely established the status quo and therefore contributed little to the policy debate. Why these BPEO recommendations should then have been rejected twice (first with the dropping of the short lived ILW component of the 1986 sites and second in 1987 when LLW was to be disposed of with ILW in deep sites) probably reflects a lack of confidence in this BPEO approach.

## More sites in 1986

The withdrawal of Billingham apparently to allow more research into the packaging and conditioning of ILW (according to the Secretary of State) and due to site access denial (according to NIREX) was followed in 1986 by three new sites for the shallow land burial of LLW and ILW wastes. The ILW component was later dropped in an attempt to appease public opinion, but it may also have reflected more detailed radiological advice that the post-surveillance period for ILW may have to be extended from 300 to 500 years; this longer period might well be regarded as involving an unfeasible committment to monitoring, maintenance and surveillance. The near surface disposal of ILW was dropped despite BPEO support for this disposal option which makes BPEO look more than dubious. Commonsense dictates that disposal of ILW in a near surface facility is putting a rather strong assumption on the containment integrity and the safety case.

This reprieve for Billingham was announced by the Secretary of State in January 1985 after a long battle by BAND, a large petition and (far more significantly) the eventual refusal of ICI to sell the mine. Without compulsory purchasing powers and with a government unwilling to interfere, there was no other option. In his statement, the Secretary of State announced that at least two further sites in addition to Elstow would be investigated to provide a comparative assessment (including an environmental assessment) of candidates as a successor to Drigg. To facilitate this comparative evaluation the government would introduce a special parliamentary development order to give limited planning permission for experimental drilling to take place. This would neutralise the attempts of local authorities to follow the steps of Bedfordshire County Council which took injunctions against NIREX to prevent them taking samples. After a long delay and much media speculation about candidate sites, the Secretary of State announced the three further sites – South Killingholme in South Humberside, Fulbeck in Lincolnshire and Bradwell in Essex – in February 1986.

*The Environment Committee's report, 1986*

This announcement came shortly after the publication of the Environment Committee's report on radioactive waste which was far from complimentary to the nuclear industry and the management of its affairs. We have already referred to much of the Environment Committee's observations and recommendations especially those pertaining to policy. Nevertheless, in March the independent review of the Best Practicable Environmental Options (BPEO) was published by the DOE. The BPEO study and the Environment Committee's report were in agreement on one major strategic issue: that long term storage offered the least attractive option of the available alternatives. The BPEO report was the outcome of a comparative assessment of disposal options as recommended by the Holliday report. After evaluating various options according to environmental and radiological risks and the associated costs involved, the report concluded that about 15 per cent of ILW should be channelled through the sea disposal route, but the longer-lived ILW would require deep underground disposal. More importantly in the context of the search for a successor to Drigg, it was argued that 'near surface disposal of LLW and some short-lived ILW is economically and radiologically attractive ... It must therefore be concluded that near-surface disposal is the BPEO for over 80% by volume of all the wastes considered' (DOE, 1986b, para 7(5)).

These conclusions were very much in harmony with the plans of NIREX and the government endorsed this strategy in response to the Environment Committee's report in July (DOE, 1986a). It reflected most of the Committee's main recommendations with regard to disposal plans, it made it clear that the sea route would be retained as a possible option but, in view of the hostile response to its siting announcement in February, decided to drop plans for disposing short-lived ILW at any of these sites despite the DOE's BPEO favourable results. The relevant quote is: 'The Government takes seriously the distinction drawn by many between the acceptability of LLW and ILW in such a site, and recognises that many people would be reassured if this site were used only for LLW' (DOE, 1986b; p 11). In essence, NIREX was now instructed to plan in the short term for LLW at one of the nominated sites while further research was conducted on finding potential sites for ILW at a deeper repository towards the end of the century.

Meanwhile, on the ground NIREX was faced with a hostile response to its exploratory drilling schedule. All the other areas more recently selected as possible sites created action groups along similar lines to that at Elstow and Billingham. Eventually NIREX had to resort to the courts, seeking injunctions to allow their contractors access to the sites. What was worrying the government about this action was that these anti-nuclear protestors were not the type portrayed by the right-wing popular press (for example, socialist hippies). These protestors were village residents, landed gentry and conservative MPs. All of the sites were in Conservative constituences, one MP (in Humberside) threatened to resign over the issue, the three others were government ministers, including the Chief Whip (in Essex).

As NIREX established public information centres in each area, circulated a newspaper (*Plaintalk*) and tried to allay the fears of local residents at public meetings, the County Councils involved formed a coalition to pool resources and information to fight the NIREX proposals. In particular, they carried out their own fact finding mission to Sweden, West Germany and France to compare the approach of these countries with that proposed for the United Kingdom. Their report, published in January 1987 by Environmental Resources Ltd, was very critical of British policy compared with that of the other countries visited. In particular, they argued that greater public acceptability had been achieved abroad because of a more open information policy and closer liaison between the nuclear industry and the local authorities. Moreover, they argued that in both Sweden and Germany, deep underground repositories had been constructed for LLW and ILW. Using cost data from the French Centre de la Manche 'near surface trench facility' and the 'deep cavity disposal' sites in West Germany and Sweden, ERL compared these costs with those quoted in the BPEO report and concluded that the cost differential between these options was much closer than implied in the DOE report.

*Sites abandoned in 1987*

Whether NIREX and the government were influenced by this line of argument it is difficult to know, but the Secretary of State announced on 1 May 1987 that the search for a near-surface repository was being abandoned in favour of a multi-purpose facility for both LLW and ILWs. The main reason for this seemingly amazing change in policy direction was attributed to costs, namely that the increase in shallow repository costs over the years had narrowed the cost differential in disposing of these wastes in a deep repository. Also, the decision to have separate disposal facilities for LLW and ILW was partly because it was hoped that the public would be considerably less concerned about LLW than ILW; actually, the potential risks are very dissimilar. However, in practice, no reduction in public hostility could be observed and this fact, together with increasingly unfavourable economics, combined to support the development of a deep multi-purpose LLW and ILW facility. With a general election very near, there was a small political advantage in making the announcement before rather than after the general election result. The feeling in the nuclear industry was that had they wished to proceed with a near-surface facility at any of the four sites, then they would have received the necessary governmental support despite the impending 1987 general election.

The increase in cost was associated with the necessity of proving a 'Rolls Royce' engineered trench to accommodate LLW (unit costs rose because of the government's insistence on excluding short-lived ILW from such a repository) and meet public perceptions rather than technical requirements. The costs quoted were £500 – £1,000 per cubic metre in a shallow repository compared with £750 – £1,200 to add the LLW to a deep disposal facility (NIREX press release, 1 May 1987). The chairman of NIREX, John Baker, argued that, 'So whereas a year or so ago when the policy of investigating the current four sites was endorsed, the ratio of cost per metre cubed of disposing of low level waste compared to shallow was probably 4 to 1, I now consider it more likely that the costs are broadly similar if low level waste is regarded as being piggy backed into a deep repository on the back of intermediate level wastes' (*Atom*, June 1987 p.24). The argument was that rather than develop two facilities (one for LLW and one for ILW) the marginal cost of adding LLW to an ILW facility was probably the same or slightly cheaper than with a separate LLW facility. It was estimated to cost between £160 – £200 million to handle non-Sellafield LLW by a special near-surface facility, compared with £140 – £200 million for piggy-backing with ILW. Baker adds 'There is not a lot of difference between costs of disposing of all low level waste in a deep repository when costed on a marginal basis compared with the average total cost of disposal in a shallow repository' (ibid.). He did however recommend that 'the option of near surface disposal for bulk decommissioning items, in particular the heat exchangers, be retained' (ibid.). This decision was obviously popular with the four targetted communities but is yet another clear example of what the Environment Committee (1986) had identified as the continuing saga of lurching from one event to the next rather than having a clear coherent strategy. On the other hand, it might be seen as a reflection of a learning curve as both NIREX and the government adapted to greater knowledge about the problems and difficulties. Opposition MPs welcomed the government's change of heart but saw the decision as a political one in view of the pending general election and the fact that three government ministers stood the possibility of losing their seats on this issue alone. There is no guarantee, however, that these sites will not be revived as possible options in the future. As greater volumes of decommissioned wastes require to be disposed of next century, the costs of cutting up, packaging and preparing the more bulky items for a deep repository could see another switch in strategy once the combined LLW-ILW development has been approved, or much earlier if the Ministry of Defence go their own way.

The announcement by the Secretary of State was apparently quite unexpected and it seemed to cause some surprise within the nuclear industry and other administrative bodies. The economic arguments were not really that compelling. What they did do was to avoid the need for two developments in the short-term rather than one. Members of RWMAC were particularly aggrieved as they were not consulted about the issue and by the end of the month at their annual meeting they had still not received adequate information from the DOE to assess the decision. One member, John Knill, engineering professor at Imperial College, maintained that the 'government had a precise strategy for the disposal of radioactive waste which the committee supported. That strategy has now disintegrated' (the *Independent*, 2 May 1987). The Committee was especially anxious that the new repository would be available before Drigg reached full capacity. Clearly, an interim strategy is to use Drigg and possibly store wastes at nuclear sites. Neither of these options have appealed to the government in the past because of reservations over the Drigg site and the cost of storing LLW. Moreover, the costing figures produced by NIREX have been met with a high degree of scepticism by some scientists, for example, Professor Knill feels that if NIREX knows enough about the costs of shallow land burial to justify abandoning the programme, it must be in a position to give detailed costings of alternative deep disposal methods (Ibid.). It is doubtful if this is the case in that it may well have underestimated the likely increase in costs of a deep disposal site that will have to be three times larger to accommodate LLW.

This change in policy direction by NIREX, endorsed by the government has already incurred considerable 'front end' costs in seeking a shallow repository. NIREX had spent £17 million in planning, evaluating, promoting and exploring this disposal route with the prospect of incurring a further £4 million in winding up the project (Parliamentary answer by William Waldegrave, 7 May 1987). In addition the DOE had spent over £1 million since 1982 commissioning research into various aspects of LLW and ILW disposal: the BPEO report cost £0.2 million. Considering that its main recommendations have been disregarded less than 14 months after its publication, the events point to an expensive change of strategy by the government or its advisers. On the other hand, the government is not responsible for NIREX and there is little doubt that in the long term much of the site exploration work will not have been wasted. When the time comes that a near surface facility is needed, then there are four well-understood sites waiting to be developed.

Attention has now focused on three main options currently considered available: a deep mine on land; a repository under the seabed accessed by tunnels from the shore; and a repository under the seabed accessed by a vertical shaft from an ocean rig or similar structure. Work has been in progress since 1985 on the first two concepts but the third option is more recent in origin. Prior to the May announcement NIREX had planned to commission such a repository by the early years of the next century (Davies, 1987, p36). These concepts and the siting criteria are discussed in detail in chapters 6 and 7.

*The way forward*

The abandonment of the LLW shallow land burial sites in May 1987 marks in many ways a major turning point in the policy story. It was, apparently, a decision made on economic grounds. The economic viability of a LLW-only radwaste dump was considered less than a combined LLW and ILW deep sited facility. It also made massive sense only to have to find one site instead of two. The government's policy did not change, it was merely that NIREX were allowed to act on their own behalf instead of being directed by the government. The strategy now is to seek a more broadly based policy based on a large scale public participation exercise designed to investigate what the public and various interested parties would find acceptable. The NIREX *Way Forward* document (NIREX, 1987) is clearly a step in the right direction. It is discussed further in chapters 4 and 8.

Finally, we would agree with RWMAC (1988a) when they point out that there are a significant number of issues on which policy remains confused or deficient. There is still enormous uncertainty about whether a deep disposal site will be found, approved, and in operation by 2005 as planned. There are doubts about the eventual disposal route for large scale decommissioning items, and no policy over ILW waste conditioning and packaging that may well pre-empt later decisions regarding storage or disposal. There is concern that LLW storage space at Drigg and Dounreay will prove inadequate before the new facility is online and fears that the potential of Sellafield as a deep dump site will be affected by the local political unpopularity of Drigg. So the *Way Forward* might be seen to be tackling only one of several inter-related issues.

**Regulatory practices**

All this policy related activity is basically designed to ensure that any radwaste dump will meet whatever radiological controls are imposed. Indeed, the government has been consistently emphatic throughout the debate about radwaste that all wastes should be disposed of under strict supervision to high standards of safety (DOE, 1986a). There has been no criticism of this policy – only how it might be translated in reality. Accordingly, the three key objectives for radioactive waste management are:

(1) all practices giving rise to radioactive wastes must be justified, that is the need for the practice must be established in terms of overall benefit;
(2) radiation exposure of individuals and the collective dose to the population arising from radioactive wastes shall be reduced to levels which are as low as reasonably achievable (ALARA), economic and social factors being taken into account; and
(3) the effective dose equivalent from all sources, excluding natural background radiation and medical procedures, to representative members of a critical group should not exceed 1 mSv in any one year; however, effective dose equivalents up to 5 mSv are permissable in some years provided that the total does not exceed 70 m Sv over a lifetime (DOE, 1986a).

These objectives have been the subject of much debate and are discussed at length by the Radioactive Waste Management Advisory Committee (RWMAC) in its annual reports of 1984 and 1985, the Layfield Report (1987 chapters 30–36) and the Environment Committee (1986) investigating radioactive waste (chapters 6–8). In short, the most recent advice from ICRP in 1985, endorsed by the NRPB and RWMAC, is that 1 mSv per annum should be the limit, instead of 5 mSv for the effective dose equivalent. However, this is to be achieved by the best practicable means along the lines of the ALARA principle. However, this involves semantics such as what is reasonable, the degree of reasonableness and the cost effectiveness of such measures. The problem is that the question of what is considered to be reasonably practicable is not easily defined but depends on cost-benefit analysis based on highly uncertain and error prone data. Yet the ALARP principle has been a feature of British life ever since the 1974 Health and Safety at Work Act. It is closely related to the ICRP's 'as low as reasonably achievable' (ALARA). The RWMAC use ALARA in respect of dose and ALARP in respect of discharges. A further difficulty is the characteristic feature of British pollution control which is based on the use of what might be viewed as 'gentlemanly agreements' rather than absolute standards as a means of defining the limits of regulatory acceptability. Provided you are doing the best you can manage, then both ALARA and ALARP might be considered satisfied.

At the Sizewell B PWR Public Inquiry there was extensive debate as to the meaning of a certain section of the 1965 Nuclear Installations Act. Section 7 (1) states that there is an absolute duty to achieve a certain standard of safety, namely that no occurrence on a licensed site involving nuclear matter causes injury to any person or damage to any property, other than the licensee's (Layfield, 1987, para.107.8). There was some debate as to whether safety assessment principles based on the ALARP principle would meet this requirement. The counter argument was that the 1965 Act did not set safety standards but merely imposed strict liability on the licensee for injury or damage in specified circumstances. Additionally, this Act provided no guidance on the level of risk considered necessary to meet stated duty and does not therefore amount to a safety standard. In other words, the Acts of Parliament only provide the broadest of contexts within which the onus appears to be on the licensee to regulate itself.

Layfield was very critical of the NII's approach to ALARA because of its inability to apply cost benefit analysis to proposed design changes to weigh up the benefits (reducing discharges) in relation to the cost. There are also fundamental difficulties in accurately assessing the 'value' to be placed on reduced discharges; these involve biological dose response models that are uncertain, putting a value on diseases that are possibly induced (partly or largely) by additional radiation exposure, and in predicting with any certainty changes in dose rates reflecting different disposal practices. The entire process is so dominated by random effects and uncertainty that it is simply unsuitable for a rational scientific approach that demands precise quantification. The result is a scientific smokescreen that attributes results which are massively uncertain and totally assumption–dependent a quality they would otherwise not possess. Under these circumstances, the use of optimisation procedures (to maximise cost benefit functions) is wholly inappropriate and potentially grossly misleading (and wrong). It has to be done 'properly' and current scientific knowledge is insufficient to underpin adequately such sophisticated technology when applied willy-nilly everywhere. Its use in radwaste decisions that can involve up to 100 million year timescales is dubious.

Nevertheless, Layfield recommended that ALARA should remain subject to a more rigorous approach being adopted. The Environment Committee went one step further by recommending that ALARA should be replaced by fixed emission standards reviewed every three years to permit more effective monitoring and enforcement, although this was rejected by the government (DOE, 1986b). The old argument still lives on: 'The Government rejects the proposal that discharge limits should be regarded as fixed standards, as this would encourage operators to regard them as targets and thus discourage them from using the best practicable means (BPM) to improve performance within the set limits. This would run counter to the well established principle of radiological protection that the doses to the public resulting from radioactive discharges should be kept as low as reasonably achievable (ALARA)' (para 66, DOE, 1986b). Much of this criticism stems from the public concern, not only at home but internationally, about the historic levels of the discharges from the Sellafield reprocessing plant. The Environment Committee (1986) were particularily vexed at the wide discrepancy between authorized discharges and the actual emissions made, in that unusual events or accidents could actually be accommodated within the certification of authorizations under the ALARA principle. Following the November 1983 leak at Sellafield, there has been much greater pressure placed on the nuclear industry to comply with stricter authorized limits. Prior to the NRPB–ICRP view that 1mSv should be the target dose limit, RWMAC recommended that a 10 per cent limit of the 5mSv level should be the waste management objective for doses to the public. Coupled with these recommendations, Sellafield was under pressure to eliminate its discharges following further adverse publicity linking these emissions with childhood cancers in the region. These allegations are still 'unproven' (Black, 1984) and despite further research there is still no explicit evidence of a radiation link. More recently, the discovery of additional cancer clusters at Dounreay and also on Tyneside (near no known radiation source) has both increased the coincidence and greatly complicated the picture (Openshaw *et al*, 1988). Nevertheless, the existence of the Seascale cancer cluster and political concern about the controversy surrounding the plant and international pressure from the Paris Commission's recommendations on the use of best available technology for reprocessing plants to prevent marine pollution from land based sources, has meant that there has been an accelerated renovation of plants at the site.

The net effect of all this pressure is that BNFL will have to invest £250 million in addition to the £490 million already spent on new plant to reduce discharges. The recent authorized investment is the Enhanced Actinide Removal Plant (EARP); an effluent treatment plant to remove alpha activity from high volume waste streams is a continuous treatment process and the removal of alpha, beta and gamma activity is a batch process for low volume streams. The plants more recently commissioned were the salt evaporator plant which allows the medium activity salt bearing effluents to be concentrated and stored instead of being discharged, Pond 5 which is a replacement storage and decanning complex for MAGNOX fuel and the Site Ion Exchange Plant (SIXEP). SIXEP treats water contaminated with radioactive wastes from irradiated fuel storage ponds by passing it through sand bed filters in exchange beds prior to discharge into the sea. By 1991 these investments plus the THORP will lead to considerable reductions in discharges such that alpha radiation emitters will be reduced to 20 curies a year in 1991 from 380 in 1983 and beta radiation emissions would fall to 8000 curies a year from 67000 during the same period. While BNFL argue that the extra £250 million investment for EARP is not cost effective in that it will only save two lives in the next 10,000 years (Atom, February 1985), the Environment Committee (1986) are quick to point out that despite these measures, Sellafield will still produce much higher discharges than elsewhere; indeed, their revised emissions are not far short of the sum of all other British nuclear discharges.

The Sellafield experience is interesting but not directly relevant to radwaste dumps except to the extent to which it demonstrates how the British style of nuclear regulation works in practice. It is also relevant as far as the public is concerned because many people would not necessarily recognize any significant distinction between Sellafield and a radwaste dump. Yet there is an argument that the essentially different nature of a radwaste dump compared with a reprocessing plant or a nuclear power station has been overlooked. They are not the same! A radwaste dump that releases any radioactive discharge direct into the environment is clearly faulty, its containment is failing and it is leaking! They should be zero-discharge facilities, unlike both reprocessing plants and reactors. Additionally, while it is fairly easy to imagine unlikely and incredible albeit feasible reactor accidents and reprocessing plant disasters, it is really difficult to imagine any plausible mechanism (other than perhaps a nuclear weapon) that could cause anything remotely devastating to a shallow land burial site, and virtually nothing other than sabotage using a carefully sited high yield nuclear device against a deep disposal facility. Quite simply, there are no short-term pathways or accident scenarios likely to cause a minor (let alone a major) radiological disaster. So why apply the same standards to radwaste dumps? It should be possible to be a few orders of magnitude more stringent because there can be no measurable external radiological impact if the facility is working properly.

## Commentary

The period from the announcement of the national strategy to its partial demise in May 1987 is marked by political intrigue as the government and NIREX progressively changed its stance to accommodate pressure from public opinion, a parliamentary committee, lobbying by pressure groups, and again the Royal Commission on Environmental Pollution. In addition, a new element entered the scene in the form of the Ministry of Defence who, now denied an easy sea dumping route for their expired nuclear submarines, started plans to establish their own ILW dump near the coast. The MOD seemed to be proceeding without much regard to the activities of NIREX or the DOE. It is in any case, more or less, exempt from the normal political and planning process.

It is also interesting that a mere eight years after the Flowers Report, the Tenth Report of the Royal Commission on Environmental Pollution felt that the effectiveness of RWMAC and NIREX would be enhanced if their membership was widened to include other areas of expertise (RCEP, 1984 paras 344 and 345). As the hostility towards land based disposal options increased, it was not surprising that the Commission recommended that local government officials would provide an invaluable input to RWMAC as they would be able effectively to anticipate the types of question that local communities might raise. Moreover, it had reservations about NIREX being able to command public support for its policies when its membership was made up entirely of persons who have a vested interest in the continuing success of the nuclear industry. Thus, it recommended that independent members with a knowledge of wider public opinion should be co-opted to the Board of NIREX. In November 1986 NIREX responded to this request and appointed a TV presenter, a prominent trade union leader and a bio-chemistry professor to the Board as independent directors. It is doubtful if this approach was the one which the RCEP had in mind. On the question of membership of RWMAC the Commission's recommendations have not been heeded, indeed both the Environment Committee (1986) and Layfield (1987) also recommended that greater environmental representation is necessary on RWMAC.

The root cause of the nuclear industry's problems is that its confidence in technical solutions to radioactive waste disposal, endorsed by RWMAC and the government is not shared by the general public whose fears of radiation are ruthlessly exploited by environmental pressure groups and the media. Some measure of the gap in understanding is seen in the 1982 White Paper which explains public fears as follows:

Radioactive waste is the source of much public concern. It is sometimes seen as dangerous and intractable material which poses almost insuperable management problems. This view is, in the government's considered judgment, an exaggerated one. Closer study of the question shows that although problems and dangers are certainly present, the problems are being resolved, and the dangers can be eliminated, by the systematic application of known technology and sound common sense. Policies to this end will not, however, be successful unless there is public support based on a full and accurate assessment of the situation. [DOE, 1982, p.19].

It was considered that it was only necessary to publish ample material for an informed public debate and all would be well. In practice, the reverse can also occur. More information and a more aware public, only greatly increase public fears. To work well, more information has to be accompanied by 'faith' in the science and scientists responsible. People who work in the nuclear industry clearly possess the necessary levels of faith and trust, most members of the public seemingly do not. If they understand the technical arguments, they see no reason to trust that the experts are right. There are far too many real and fictitious examples of where the experts get things wrong. The naïve often see things in a very different light than that of the experts; for example, a barrister's questions put to a nuclear scientist are often extremely penetrating and politicians operate in this maximal use of minimal knowledge mode. The solution, if there is indeed one, is not going to be easy; indeed it may (with retrospect) have been a major mistake to ever consider that public acceptability was important. There are undoubtedly a small number of areas where even in democracies public debate has been suppressed or where major irreversible decisions have to be made without a large popular mandate. Perhaps, radioactive waste management is one of these and the nuclear industry, so full of their own self-importance and confidence, wrongly assumed that its popularity extended beyond the confines of the nuclear industry and gambled that provided all the democratic and legal niceties were observed, all would eventually be well.

The use of the standard 'fixes' to legitimize public policy initiatives seem to have little effect when applied to nuclear waste. For example, the approach of building safer facilities or modernizing plant to minimize discharges has not reduced public anxiety about Sellafield. Administrative fixes, such as the creation of RWMAC, NIREX and a DOE waste management division, have not solved the problem and the behavioural fix approach of bombarding the public with PR campaigns as typified by UKAEA, BNFL and NIREX has, if anything, made the layman more cynical of the nuclear industry's motives.

A research team investigating the Sizewell B Public Inquiry argued that 'radioactive waste disposal policy had been subjected to considerable administrative and political pressures resulting in policy flux and ambiguity' (Kemp, *et al*, 1986, p.22). They claim that the quasi-administrative, quasi-judicial process of the public inquiry exposed the contradicting demands placed on the government to resolve the waste disposal issue. Thus, conventional policy making models would not provide an adequate explanation of state policy formulation in an issue as complex as this one. For example, it can be argued that nuclear power policy in general in the United Kingdom had until the 1980s been loosely based on dualist models of state policy formulation. Here, the internal influences upon policy development (the relationship between DOE and Department of Energy) and the responses of policy making bodies to external influences (mainly pressure groups) had produced an incremental policy of steady nuclear development. Nuclear waste, however, is a different matter. The response externally has been primarily negative with all political parties at one level uniting in their opposition to NIREX's proposals. Clearly, some of the positive benefits associated with other nuclear facilities are not associated with waste disposal depositories.

This complex situation has been responsible for the government's ambivalent stance towards radioactive waste disposal. It is therefore surprising that Layfield in one of the weakest and thinnest chapters of his report should comment that 'since 1976 a clear strategy for the management of radioactive wastes has been established by government,' (chapter 41 para.51). This was not the view expressed by Kemp *et al* (1986) or the Environment Committee(1986) which investigated radioactive waste issues in 1985. Indeed, the Committee comments on policy issues that: 'inconsistences and the erratic *ad hoc* way in which they have emerged from government and industry demonstrate just how seriously the UK lacks an underlying, coherent strategy' (para.250).

This is of particular relevance to the government's siting guidelines which went through three phases – a consultative document coinciding with the Secretary of State's announcement of the Elstow and Billingham sites in October 1983; an advanced draft presented at the Sizewell B Inquiry in September 1984; and the definitive document coinciding with the Secretary of State's announcement of his revised plans of January 1985. Thus, NIREX was at an advanced stage of site evaluation before the consultative exercise had been published. We argue that this is again the case with the *Way Forward*. It is interesting to note the subsequent changes in the subtle wording of key siting guidelines as they moved to the final version. Clearly, major efforts were made by the DOE to avoid statements conflicting with those of the developer, NIREX. For example in their evaluation of alternative sites, the developer 'will be expected to show that he has followed a rational procedure for site identification'. However, the guidelines continue 'he will not be expected to show that his proposals represent the best choice from all conceivably possible sites. Where factors other than radiological ones have influenced the choice of site, this must be brought out clearly' (DOE, 1985, p.19). It is not clear as to why NIREX did not appear to conform to these principles for the Elstow and Billingham sites since they would have been aware of the draft DOE guidelines.

It is also interesting that the final version of these guidelines have a major difference compared with the version submitted as evidence to the Sizewell Inquiry in 1984. The advanced draft document stated 'that in selecting his preferred site, he has not ignored a better option for limiting radiological risks' (DOE, 1984a, p.14). Moreover, the original 1983 consultative version of the same guidelines also added 'or for satisfying town and country planning criteria and policies'. These changes amount to a deliberate de-emphasis on the importance of the site comparisons and the evaluation of a wider range of alternatives. Presumably they reflect NIREX's comments on the earlier draft. It would certainly have increased the complexity of their task but it would also seem to have been quite reasonable criteria from a public acceptability point of view. Obviously, there are limits to what the DOE can demand from NIREX, but this particular area of major de-emphasis does have an impact on the entire site search and evaluation process. Clearly NIREX should be pushed as hard and as far as possible in their search for 'optimal' sites. It is important that a demonstrably nearly optimal site should be selected as the overall aim should be the best practicable environmental option. Obviously it will in practice be impossible to prove that 'the best' sites have been found but it should be relatively easy to demonstrate that an exceptionally good site has been found with little prospect that a significantly better site exists. Both require a comprehensive search and evaluation process and there is some concern that NIREX has not gone far enough to be able with any great confidence to be certain that they have found a 'good' site; see chapter 7. Ironically, this nearly optimal assurance would have been considerably more feasible had NIREX been granted compulsory purchasing powers.

The Environment Committee (1986) take this line of reasoning further by claiming that NIREX were identifying specific sites when not only the siting criteria but the policy parameters had not been properly identified by the review exercises then underway, for example the Holliday investigation, Best Practicable Environmental Options, in addition to the aforementioned consultation exercise. In conclusion, they argued that this uncoordinated approach was the price paid for the loss of Billingham as a site for long-lived ILW disposal. Meanwhile, RWMAC (1988b) talk of the need for contingency planning in case the current siting attempt fails. The added uncertainty likely to be caused by privatization merely adds another random effect to the solution equation.

# 4

# The NIREX story

## Some background issues

At this stage it is useful to examine the creation of NIREX (Nuclear Industry Radioactive Waste Executive). NIREX is a private company established in 1982 by the nuclear industry (BNFL, CEGB, SSEB, and UKAEA) to 'solve' the civilian radwaste problem in a speedy and efficient manner. It should be remembered that the 1960 Radioactive Substances Act gave responsibility for the disposing of waste to the nuclear industry that created it. This 'polluter must pay' principle is an accepted part of the government's environmental strategy. So it is hardly surprising that waste disposal was seen as a normal commercial activity that could presumably pay for itself and it was thought it should be handled in the normal way like any other industrial development. The nuclear industry has long harboured this fundamental mis-apprehension that it is just like any other industrial enterprise. It was thought that as no one was likely to be harmed by a radwaste dump, so there was no need for NIREX to be given special powers; for instance, it does not have the compulsory purchase powers of the UK Atomic Energy Authority, nor is it a statutory undertaker and thus exempt (like the CEGB and SSEB) from the Town and Country Planning Acts. The history of radwaste disposal would have been very different had the task been given to the UKAEA to solve in the 1960s or even the 1970s. Additionally, government felt obliged to establish a regulatory framework within which NIREX has to operate. The 'rules' cannot of course be too stringent in case NIREX defaults or is frightened-off and the government has to take over all responsibility itself. At the same time, the rules cannot be too lax because the government has to be seen as an independent authorizing body that conforms with various international conventions and has deliberately sought to distance itself from NIREX. NIREX has enjoyed strong governmental support and it views itself as an instrument of government radwaste policy, but it is a separate business with the right of self-determination.

In this review of NIREX it should also be remembered that NIREX is only one of two potential radwaste developers in the United Kingdom and that currently NIREX has no responsibility for the heat-generating high level wastes. The Ministry of Defence (MOD) also has a vested interest in the disposal of radioactive wastes, but unlike NIREX the MOD has a number of major advantages: (a) it can be exempt from the various Acts of Parliament relating to radioactivity; (b) it is responsibly for its own safety; and (c) in the interests of national security it is not open to public scrutiny nor concerned with public opinions or debate. It may be imagined that the reprocessing wastes from the manufacture of nuclear weapons are stored together with civilian wastes at Sellafield and that the combined processing of both civilian and military fuel successfully hides the military's contribution to the nations' LLW and ILW stocks. Furthermore, there are other low level dumps other than Drigg for the disposal of slightly radioactive materials. However, the MOD's principal future needs seem likely to be the highly visible task of decommissioning nuclear submarines. The larger pieces are probably either too large for disposal by the planned NIREX deep sited facility or they may present unacceptable operational risks. A similar problem will occur with the third stage decommissioning of nuclear power stations, except that the timescales are vastly different; power station dismantlement will probably not occur for at least 100 years. The MOD, however, is facing its problem now. It is probable that additional near surface disposal facilities will need to be developed sometime fairly soon to handle these bulky components and it is not clear as to whether the MOD will simply proceed on their own account and develop (in secret?) their own facilities, or whether they will wait for the outcome of the current NIREX effort.

It is important to have some sympathy for the enormity of the task that NIREX faces and to appreciate that many of the problems have been created by the civilian nuclear industry itself in its attempt to 'play fair by the rules'. Whether this was a deliberate and conscious act is difficult to determine as it is not clear as to whether there was a real alternative strategy. However, it was certainly brave. Perhaps, in the early 1980s someone, somewhere, made a fundamental mis-assessment of the public and political realities of the radwaste dump problem by grossly underestimating public sensitivities to the subject. On the other hand, this public responsibility route is also being followed in other countries; for example, in the United States; and it will in the end succeed; indeed it can be argued that in the longer term it is the only satisfactory solution. Public acceptability both now and in the future are essential pre-requisities for any successful radwaste dumping strategy and this single, simple factor more than anything else will determine the long term success or failure of any proposals. Ultimately, the existence of some kind of facility is not in doubt because, like it or not, there has to be one. The real question is where to locate Radwaste Dump No. 1; the first of a sequence of similar facilities. The subsequent dumps will be far more easily sited once the initial largely psychological barrier is breached. However, it is also argued that initial success in building a facility, even with construction costs of £200 – £800 million, is no guarantee that it will survive more than one parliamentary session. Unlike nuclear power stations which once built are likely to be used, radwaste dumps are far more cancellable entities. Permission to construct one is not sufficient if building it creates so much local opposition that cancellation is a politically attractive option at some future date. It is thought that NIREX has only recently appreciated these finer political points that are of such tremendous longer term importance to their future operations. The importance of these general background political aspects recur on several occasions during this historical review of NIREX and its previous siting attempts.

Before examining the details of recent history, it is worth recalling that the early 1980s were perceived to be a time of crisis with respect to the need to establish a radwaste disposal facility. The Flowers Report (1976) had drawn attention to the problem and the government was starting to react to it. Perhaps also the nuclear industry was starting to panic. There was clearly a need to establish a facility of some kind in order to justify the extensive development of reprocessing capacity at Sellafield (namely THORP), to avoid criticisms associated with the planned development of nuclear power stations that no disposal facilities existed, and to prepare for the golden age of the fast breeder reactor. The absence of any long term acceptable disposal route for low and intermediate level wastes was clearly a growing embarrassment.

The disposal of low-level solid wastes is currently carried out at the shallow land burial site operated by BNFL at Drigg in Cumbria and at Dounreay by UKAEA, while very low level wastes (VLLW) are disposed of at 40 or so landfill sites scattered around Britain. There is some concern that this VLLW dumping is not being properly managed, that there is no long term site surveillance and probably no records of what precisely has been dumped, and there is no mechanism to avoid the subsequent redevelopment of former VLLW dumps for incompatible uses. It is also doubtful whether the public know where these VLLW dumps are located, despite a requirement for Local Authority participation. It is clearly much better to centralize the disposal of all kinds of LLW.

Currently, Drigg is the place where most is dumped. Drigg might be regarded as the archetypal British LLW dump. Here the unpackaged waste is dropped into boulder clay trenches eight metres deep and covered with two metres of soil. It was this approach that the Environment Committee complained about in their 1986 report; they wrote, 'We conclude that Drigg is not an acceptable model for any future disposal site' (p. xxxvii). However, the criticism was accepted and this image is now changing with the development of engineered vaults filled with containerized waste to replace the old trench disposal system. The old trenches are also being capped against rainwater infiltration. However, it is important to note that Drigg is operated by BNFL and not NIREX. Currently, NIREX do not have any radwaste dumping facilities of their own. For a brief time NIREX were given the task of running the sea disposal route. About 15,000 cubic metres (1982 figures) of packaged wastes were annually being disposed of in the north east Atlantic. However, action by the National Union of Seamen, environmental groups and international treaty effectively removed this option in 1983. The government finally agreed in 1988 to abandon sea dumping as a ILW option while still wishing to retain it as a possible means of disposing of large items of decommissioning waste.

Since the Drigg site was opened in 1949 it has accommodated about 500,000 cubic metres of low level wastes (mostly from BNFL), but if no other site(s) are used for future waste arisings Drigg will be full soon after the turn of the century. As Drigg is essentially part of the BNFL reprocessing works there is a strong argument for reserving Drigg exclusively for Sellafield wastes and opening a new site elsewhere in the United Kingdom for the 170,000 cubic metres of low level wastes projected to arise from other sites (power stations, research establishments, industry, hospitals etc) by the year 2000. This would allow the Drigg site to serve BNFL well into the next century. There was also a need to develop deep disposal facilities able to accept intermediate level wastes not suitable for disposal at Drigg. These wastes are currently stored at nuclear power stations, at some UKAEA sites and at BNFL Sellafield. A projected 99,770 cubic metres of intermediate level waste were expected to arise by 2000 and it was government policy that this should be disposed of as soon as possible to avoid the need for costly and extensive storage facilities. It was this twin set of needs that resulted in the creation of NIREX.

## NIREX: its set up, organization and authorization

NIREX was created with government approval in 1982 by the UKAEA, CEGB, SSEB and BNFL (the four main bodies in Britain's nuclear industry) with the aim of creating a national body to implement a coherent overall strategy for the safe management and disposal of low and intermediate level wastes. An announcement to this effect was made in a White Paper, *Radioactive Waste Management* (Cmnd 8607, 1982). In it the government attached the 'highest importance to the safe management of radioactive wastes' and expressed confidence that the safe and acceptable disposal of wastes is an achievable goal. The Flowers Report (1976) had previously recommended that a nuclear industry independent body (the Nuclear Waste Disposal Corporation) should be created to develop and manage disposal facilities. However, this advice was rejected by the government and NIREX was eventually created instead. The White Paper explains the decision as follows: 'the government has agreed that the component parts of the industry should, in co-operation, set up a Nuclear Industry Radioactive Waste Executive (NIREX) in order to provide a mechanism by which they can successfully fulfil their own responsibilities in this field and work within a comprehensive plan for waste management' (para 57, p.17). There are some hints here that maybe the government might have preferred a different approach but was persuaded otherwise. One factor was undoubtedly the cost of developing such a facility, which was then estimated at about £65 million spread over 10 years (para.59, p.18). Another was the involvement of the private sector in an area that hitherto would have certainly been regarded as a matter for government. It also made sense to have the principal nuclear industries working together for a common solution. Another argument in favour was the problem that would arise if the same government departments were responsible for: (a) devising a national strategy, (b) setting safety standards, (c) giving themselves planning permission, (d) building a disposal facility, and (e) monitoring their own successes. It was far easier to appoint some other agency to perform the task so that at least a nominal degree of separateness and independence would appear to exist. Finally, it was declared that 'Proposals for new facilities will also be subject to the normal planning legislation, including the provision for a public inquiry' (para 62, p.18) thereby ensuring legitimation.

The main task of NIREX was to handle the disposal of low and intermediate level wastes. NIREX consists of an executive unit staffed by UKAEA staff which is controlled and supervised by a directorate made up of senior representatives of the partnership organizations (see Figure 4.1). NIREX is funded entirely by its component bodies. According to The Third Annual Report on the work of NIREX, 'The NIREX Executive Unit is staffed by scientists, engineers and other professional and support staff from the UKAEA and CEGB; the Unit carries out the day-to-day work of NIREX' ( para 3, 1985). Most of the detailed work seems to be subcontracted out. The Board of NIREX originally did not contain any independent members and its proposals are scrutinized by the Radioactive Waste Management Advisory Committee (RWMAC) which has been set-up to advise the government on waste management issues. It is important to note that NIREX's responsibilities do not cover the storage and disposal of high level or heat generating wastes. These wastes are not considered to be a matter that requires an early solution. They are currently in storage at Sellafield awaiting vitrification and ultimate disposal. Finally, it should be remembered that NIREX was given no special, extra-legal powers. It has to operate like any other commercial enterprise and is subject to regulation by a number of government departments and regulatory bodies.

In November 1985 UK NIREX Ltd came into being. NIREX was now no longer a division of the UKAEA but a legal entity able to operate in its own right in a commercial environment. Previously NIREX was not legally allowed to employ consultants, to take on contractors or even to purchase land. The customers are primarily the shareholders so that UK NIREX Ltd is seemingly not in the business of making money out of waste, although no doubt this is a possibility once there is a facility in place. Indeed, there are no legal restrictions precluding NIREX from seeking customers on a world-wide basis. This may be seen to be somewhat in conflict with the national interest justification that will be used to obtain approval for any radwaste dumps in Britain, but at present it is purely a theoretical consideration. The media and political fallout is obvious if Britain was to be seen to become the nuclear dustbin of Europe or the world and no doubt some thought may need to be given to foreclose this future option for an aggressively commercial UK NIREX Ltd. At present, there is no service they can sell.

Clearly NIREX are servants of political masters. NIREX views itself as following government policy on nuclear waste disposal. As and when new policies are made, NIREX will conform to them (*Plain Talk*, November 1986, p.2). However, after the privatization of the electricity industry, the government may no longer be able to control NIREX to the same extent; the Department of Energy holds one share. As a commercial organisation NIREX could presumably seek to continue operating its radwaste facilities regardless of any changes in government policy provided they met current safety standards. What is important here, however, is that the timescale for the development of any major radwaste dump would now be such that the continued support of two, maybe three, different governments might be necessary before operations could start. It would remain vulnerable, therefore, to any strong politically backed calls for closure or abandonment and the price of closure would be quite affordable. To this extent, the NIREX preferred dump facility will remain vulnerable to political interference for at least the first 30 – 50 years of its existence; after that, it could well be the basis for big and highly profitable business.

## NIREX proposals for radwaste disposal (1983–5)

In October 1983 the Secretary of State for the Environment, Patrick Jenkin, was able to announce in Parliament that NIREX after 14 months of extensive desk studies had identified two sites as being potentially suitable as low and intermediate level waste repositories. These were the CEGB's Elstow storage depot in Bedfordshire, and a disused anhydrite mine owned by ICI in Billingham, Cleveland. This announcement coincided with the publication of the Department of the Environment's draft assessment principles (DOE, 1983) which laid out the process of site evaluation and preceeded the publication of the government's general strategy for radioactive waste management (DOE, 1984). The apparent haste reflected the desire to have a disposal facility in operation by the end of the 1980s; indeed, the Secretary of State for the Environment announced in October 1983 that, 'There is need to bring into operation by the end of the decade land disposal facilities for intermediate level wastes' (*PlainTalk*, November 1983, p.1). Both these sites were considered as potentially suitable candidates although further on site investigatory works would be required before any decision to proceed with proposals for either site.

The Billingham site is a disused anhydrite mine with an average depth of 140 – 280 metres. It is owned by ICI who were prepared to consider its use as a radioactive waste dump. The Elstow site is owned by the CEGB and consists of old worked-out clay pits which are gradually being filled up with domestic rubbish from London. This site was to be investigated to determine its suitability for the disposal of short-lived intermediate and low level wastes, while the longer-lived intermediate and low level wastes would go to Billingham. Further details are provided in chapter 5.

## Public reaction

The principal difficulty with the Billingham site was that the mine lies under an urban area. Billingham is a town of 35,000 people and part of the mine lies beneath a large housing estate. Table 4.1 gives the population in 1981 for various distances around the entrance to the mine. It is estimated that 113,963 people live within five kilometres of the current mine entrance, many of these people living directly over the workings. No wonder the proposal was not locally popular! However, it was argued (not without justification) that this was irrelevant if nobody was going to be affected by the development; for instance, if the background level of radiation was not significantly altered and the landscape was unchanged. The problem is that of explaining to these people that the development consitutes no risk or danger to them and this is clearly a virtually impossible task for a number of reasons: (1) the safety assessment resorts to probability arguments that are inherently difficult for the ordinary person to understand; (2) the engineering design was still at the conceptual stage; (3) there are no examples anywhere else in the world of siting a radwaste dump under people; and (4) fear and neurosis cannot be easily dispelled by scientific arguments, propaganda, and public debate when so many people feel threatened. It is hardly surprising, therefore, at a time when the public are no longer apathetic to nuclear developments, and have been sensitized by the media, that from the very outset NIREX's proposals were extremely unpopular. The Teesside area already has about 15 per cent of the United Kingdom's registered hazardous industries with chemical complexes at ICI's Wilton and Billingham plants and large petrochemical plants along the River Tees. This has been tolerated because it brings employment and wealth for the Teesside community but radioactive waste is 'unwanted and quite alien'. Moreover, it offers no economic benefits other than about 100 jobs.

*Table 4.1* Population around the proposed Billingham site

| Distance band | Total population | |
| --- | --- | --- |
| | 1981 | 1971 |
| 0 – 1 km | 0 | 1,535 |
| 0 – 2 km | 10,100 | 10,226 |
| 0 – 3 km | 32,266 | 38,879 |
| 0 – 4 km | 75,182 | 87,389 |
| 0 – 5 km | 113,963 | 127,751 |

Source: Calculations based on 1981 and 1971 small area census data

In their statement on their position in relation to the proposals ICI made it clear that whatever the findings of NIREX, they had no wish to pursue a proposition which was unrelated to their traditional business activities. ICI conceded they would reconsider if it was shown that the development of the mine was in the national interest, if it was the best site available and that the proposition was safe (ICI, 1983). In order to help NIREX evaluate the site's potential ICI made available all its geological reports on the mine which provided NIREX with more information than could have been obtained from a five year drilling programme at a greenfield site (NEI, 1985).

Immediately after the site announcement in October 1983, BAND (Billingham Against Nuclear Dumping) a community pressure group was formed to fight the NIREX proposals. BAND succeeded in gaining much local support from members of the public, trade unions, councils, some local industry, conservation groups and all the local MPs regardless of their political party. Frank Cook (MP for Stockton North), in whose constituency the Billingham mine lies, has been particularly vocal in parliamentary debates on the subject of radioactive waste disposal. In October 1983, he expressed his concern over the disposal of ILW at Billingham which he considered important because of the dense local population and the high concentration of volatile industries in the nearby area (Hansard, 1983). In November 1983, Mr Cook again expressed his concern and suggested that the proposals were causing such public concern that they were having a 'blighting effect' and causing property values in the area to drop dramatically (Hansard, 1983).

As a result of the 'broadly based' concern being shown by the people of the Billingham area, ICI issued a second press release in which they stated:

It is clear that local opinion is intensely opposed to the mine being used for the storage of radioactive waste. Because ICI greatly values its relationships with its employees and the community, on which its business depends, it has taken careful note of all these reactions … ICI recognises that decisions in the national interest can be taken only by Government and the appropriate public authorities as the elected representative institutions. But having carefully considered the implications that the proposal would have for its business ICI has concluded that the proposal would not be in the company's best interests and is therefore opposed to it. [ICI, 5 March 1984]

On 4 May 1984, Mr Cook presented a petition to the House of Commons on behalf of the residents of Cleveland with 83,000 signatories in support of calls to abandon the proposals for using the Billingham mine for ILW disposal. In justification, the letter of petition stated that the residents of Cleveland believe

that research into the safety aspects of nuclear waste storage has been inadequate and that there is insufficient evidence of guaranteed safety for the general public, that it is completely irresponsible to dispose of such hazardous material in or near to an area of high population, that the population of Cleveland is already vulnerable because of the high concentration of petrochemical and pollutant industries nearby … that all the efforts of recent years to improve the local environment, to improve standards for the local people ... will be to no avail if a decision is taken to store nuclear waste here. [Hansard, 1984]

The prime minister's response is interesting because she stressed that 'NIREX is not part of government but was established by the nuclear and electricity generating industries to develop and maintain disposal facilities for low and intermediate level wastes' (letter to Frank Cook, MP dated 24 May published in *PlainTalk*; 1984, p.3). NIREX see themselves somewhat differently. The NIREX chairman is reported in 1983 as saying: 'The function and position of NIREX are fairly simple. It is an executive body, and it was set up to give practical expression to government policy regarding the safe storage and disposal of radioactive waste. NIREX does not originate or determine public policy – we execute it ... It is then of course for the Government to say whether our plans are good enough' (*PlainTalk*, November 1983, p.4). Clearly the politicians have a simple means of distancing themselves from NIREX if necessary.

In January 1985 ICI withdrew its consent and cooperation and NIREX was 'invited' by the government through the Secretary of State for the Environment, Patrick Jenkin, not to proceed further with its proposals so as to avoid any uncertainty at Billingham while research was carried out into the improved conditioning of ILWs. In his statement Mr Jenkin denied that there was any suggestion of the Billingham site being abandoned because it was regarded as being unsafe in any way. In reality, once ICI decided not to sell the site, then NIREX could no longer consider it available as a radwaste dump because they do not possess powers of compulsory purchase.

The Billingham anhydrite mine remains possibly the best site in the UK for a deep repository for ILWs given its apparently excellent geological and hydrogeological characteristics, its ready to use situation and its accessibility to the nation's main source of ILW at Sellafield. The Environment Committee (1986) use a quote from William Waldegrave to represent their feelings about the absurdity of the loss of Billingham. Waldegrave said: 'I have never met a scientist interested in this subject who did not say to me that Billingham was the most magnificent site in terms of scale, depth, geology, and so forth' (Environment Committee, 1986; p.cx). The loss of this site was clearly a great disappointment to both NIREX and RWMAC, removing the natural first choice in their search for a deep repository for ILW. In terms of population density and proximity to hazardous industries however, the Billingham site was a major error. Commonsense should have ruled it out of consideration and it is perhaps a reflection of NIREX's initial enthusiasm and gross over-confidence that it ever wished to develop it in the first place. NIREX had not counted on the strength of the local public and political pressure against their proposals and had to concede that despite all the geological and economic evidence in its favour the Billingham site remains unsuitable simply because it is not remote even though there are no scientific grounds for believing that it would be unsafe. This later statement has recently been contradicted by RWMAC

(1988b) when they express some concern about the thickness of the anhydrite deposits, its proximity to acquifers and the possible effects of adjacent unfilled mine workings. However, it is very likely that the Billingham site will again be considered at some future date when perhaps public perceptions of the 'dangers' will have been reduced by either greater knowledge and experience, or (more likely) by large scale bribery in the form of large financial inducements to the local community.

Table 4.2  Population around the proposed Elstow site

| Distance band | Total population | |
| --- | --- | --- |
| | 1981 | 1971 |
| 0 – 1 km | 0 | 150 |
| 0 – 2 km | 1,098 | 692 |
| 0 – 3 km | 5,572 | 8,721 |
| 0 – 4 km | 27,039 | 36,850 |
| 0 – 5 km | 44,540 | 55,349 |

Source: Calculations based on 1981 and 1971 small area census data

The Elstow site was also a location of considerable convenience to NIREX. This time the site owner was not a problem but there was an urban area nearby; Table 4.2 shows the 1981 population distribution around the site. The levels are much lower than around Billingham but it is also clear that this site was by no means sparsely populated. Some 44,540 people lived within five kilometres in 1981 and many of these residents would probably feel themselves threatened by the development. Certainly, if there was a remote but credible accident (say, a large fire) then they might even face some small increase in cancer risks. Again it would be a 'brave' and 'confident' industry who would choose to build their first radwaste disposal facility in such a location. It would have been so much easier to select a location with fewer people close at hand. These are after all the likely future victims of a whole host of media inspired scare stories and no community could possibly be thrilled at that prospect or at the thought of being a neighbour to such a large concentration of radionuclides. Additionally, the wastes were only to be relatively short-lived and would have decayed away to negligible levels of radioactivity within 200 – 300 years. Yet Elstow occasioned another large-scale public reaction. The reaction here was more hysterical and less easy to understand than was, perhaps, the case with Billingham. Elstow then was the first of the 'not in my backyard' (NIMBY) sites where the principal objection is to the concept of a radwaste dump in an area near to where people live. Their fears and concerns are no less real than at Billingham where they were seemingly just as difficult for NIREX to appreciate. However, public opposition is now the status quo of the nuclear business so perhaps it should not have been too surprising. Indeed, it is now hard to imagine any new nuclear related development that will not arouse probably massive public opposition.

The somewhat high-handed and arrogant manner by which NIREX, in common with the nuclear industry in general, has tended to project how it operates does nothing to diminish the intensity of local feelings. For instance, NIREX (1985) is probably quite correct in pointing out that, 'The problem of having to satisfy both a national need and local concern is not uncommon; it arises in the siting of many major facilities such as coal mines, power stations, airports, and motorways. It is for this reason that the ultimate decisions on the siting of such facilities are taken at the national rather than the local level,' (pp.10–11). The problem is that the public do not seem to consider that a radwaste dump is in the same category of development as a motorway or airport and that local interests should be so easily ignored or over-ruled. It also emphasizes the importance of carrying the debate to Westminster. The traditional 'British approach' to unpopular developments is to seek legitimization via the planning process with a Public Inquiry held by an independent inspector as a means of defusing the situation; and with licensing by a collection of seemingly independent, neutral, and unrelated government departments (for example Nuclear Installations Inspectorate of the Health and Safety Executive, the Department of the Environment, and the Ministry of Agriculture, Fisheries, and Food). However, the result of any nuclear development can really seldom be in doubt once it reaches the planning inquiry and licensing stage, ultimately it is the cabinet and not the inspector and not the individual government departments who make the final decision. An inherent pro-nuclear bias is undoubtedly built-into the system but the task is made easier by the great care that NIREX would exercise in preparing their case. There would undoubtedly be substantial costs involved because no government would be prepared to fall on the radwaste dump issue and the result would be a lengthy delay while all avenues of opposition were considered.

So if NIREX really wants an Elstow development they will eventually succeed in attaining approval for it, but this 'fact of life' in no way reduces their responsibilities to explain why Elstow as distinct from anywhere else. They may consider themselves to be an ordinary industrial and commercial enterprise and as such have no extra responsibility to be especially public sensitive. Sadly, the nuclear industry has been slow to appreciate that public participation in the decision making process is important and this is not to be confused with public information campaigns. 'Open door' policies are to be welcomed. Propaganda and public education campaigns, public meetings, information centres and media seeding are all useful devices to raise the level of awareness and debate. But there are no unique set of 'facts' that can be immediately recognized as the 'truth' that will instantly dispell rumour, false fears and unjustified concerns. Nuclear neuroses are simply too strong and too easily re-created for such a simplistic strategy. You could probably steamroller through unpopular proposals but with radwaste dumps it is not a secure basis for the future. Commonsense suggests that you seek to minimize and avoid sites likely to engender massive public opposition. The siting process cannot be a purely technical exercise but has to incorporate within the technical constraints an explicit procedure for the minimization of public fears and opposition. An effective strategy is simply to avoid areas of massive local opposition. Billingham and Elstow effectively demonstrated the futility of trying to apply to radwaste dumps the same approach that was previously characteristic of nuclear power station siting.

The radwaste dump business is really fundamentally different from other large scale national interest developments and requires a fundamentally different approach to that taken when siting, for instance, an airport. Care, compassion, thoughtful statements, humility, and a slow step by step infiltration of the local consciousness, are all required; and even then success is not guaranteed. Merely because NIREX owns a potentially suitable site and no off-site consequences are anticipated under any circumstances, this is not sufficient in this game. If you cannot carry at least a large minority of local public opinion then maybe the site is undevelopable. Sooner or later, it will be closed because of political opposition. The money spent will be wasted and the goal of having an operational facility unsatisfied. These basic lessons were sadly not properly learnt from the experiences with these first round sites.

## Round two site proposals (1985–7)

On 24 January 1985 Patrick Jenkin, Secretary of State for the Environment, announced in Parliament that plans to use the Billingham anhydrite mine for ILW disposal had been dropped. This left NIREX with only Elstow as a candidate site for a radioactive waste repository. Mr Jenkin therefore decided that NIREX should find a further three alternative near surface repository sites to be examined alongside that at Elstow. This requirement was in fact present in the original draft of the DOE's siting principles published in 1983 at the same time as the two original sites were announced. Another change was the realization that there was no longer the same pressing urgency to develop a radwaste dump by the end of the decade. NIREX was now suggesting that in the short term there was little to choose from a technical and economic viewpoint between continued storage and disposal (*PlainTalk*, September 1985, p.1). As such, the lack of a disposal facility would no longer prejudice the future of the British nuclear industry, since storage was feasible until disposal facilities were available. The chairman of NIREX declared his belief that there were no technical problems in disposing of waste and that it could be achieved comfortably within current technology.

As a reaction to the opposition that followed the 1983 siting proposals, it was decided that planning permission for the test drillings at the proposed new sites would be granted by Special Development Order (SDO). Normally such investigations would require planning permission, but given the sensitivity of the proposals this would have only ended in a lengthy public inquiry and thus in view of the urgency to find a new LLW site before the end of the century it was decided to issue the field investigations with a SDO so as to allow work to proceed immediately. These 'limited permissions' were given only for the geological investigations and carried no presumption that permission would be granted for the actual development of the site. Before the development of any disposal site could take place, then, now or in the future a full scale public inquiry would have to be held to examine the alternative sites and the environmental impact they would have. However, the SDO steamroller served to greatly inflame local attitudes to the possible radwaste dump development.

It was over a year later on 25 February 1986 that NIREX announced four sites for investigation as near surface repositories for short-lived intermediate and low-level wastes (NIREX, 1986a). These included two inland sites, at Elstow and Fulbeck in Lincolnshire, and two coastal sites, at Bradwell in Essex and South Killingholme in Humberside. The short-lived ILW component was later excluded in an attempt to reduce the expected public opposition. The subsequent geological investigations cost almost £10 million and were necessary because of the sparsity of geological information about clay as a containment for a radwaste dump and to permit the radiological impact of the proposed developments to be calculated for each site. The field studies involved drilling hundreds of holes and digging pits to provide a comprehensive understanding of the geological make-up of the sites. The studies investigated the depth and consistency of the clay, its chemical and physical properties, and particular attention was given to groundwater and geological factors associated with groundwater movement. It was intended that should any of the sites prove to be unsuitable then they would be excluded. Those that were suitable would be subjected to detailed comparison and assessment. NIREX was expecting eventually to recommend one site for development. Studies were also made of the ecology of the areas so that appropriate conservation measures might be identified. In May 1986 the SDOs on the four candidate near surface repository sites were passed through Parliament and field investigations began in mid-July leading to more than nine months of protest, strained local relations and court injunctions. The work continued until May 1987 when the decision was made to abandon the concept of a near-surface repository for LLW in favour of deep disposal of both LLW and ILW.

**Political problems**

A new complication appeared in 1986 when NIREX was attacked from within the government system. A report by the House of Commons Environment Committee on Radioactive Waste published on 12 March 1986 was critical of the way radioactive waste disposal was being managed in the United Kingdom (Environment Committee, 1986). It highlighted the short-falls in the standards of operation at the Drigg LLW disposal site and pointed to its limited remaining life, whilst new revised estimates of the amount of LLW arising by the turn of the century suggested a 30 per cent increase on previous estimates, adding further pressure to find an alternative site to Drigg. The committee also, more importantly, recommended that any future near surface repository should only accept LLW and all ILW (both long and short lived) should go to deep underground repositories.

On 14 March 1986, only two days after the Commons Environment Committee Report, the DOE published its own assessment of the management of LLW and ILW (DOE, 1986b). In this report the DOE declared that the best practicable environmental option (BPEO) for LLW and short-lived ILW is near surface disposal in appropriately designed trenches; shallow land burial (SLB) for LLW and engineered trench disposal (ETD) for ILW. These DOE recommendations contradicted the Commons Environment Committee's suggestions. Perhaps surprisingly the government acceded to the view of the Commons Environment Committee and told NIREX that only LLW should go to a near surface repository. The hope was that this cautious move would help reduce public opposition, which it failed to do because the public still did not understand the difference between LLW and short-lived ILW or consider that the distinction was important. This action also greatly increased costs and it was this policy modification which eventually led to the abandonment of the near-surface disposal concept altogether in May 1987.

## Further public opposition

Worse still, the public opposition to the NIREX proposals was greatly increased by the SDOs and showed signs of becoming institutionalized. To succeed in the longer term, under perhaps different governments, NIREX does require all party support. By 1986, this was seemingly an impossible goal. They obviously enjoyed the support of the Conservative government but this is not sufficient because of the long lead times involved in repository building and given their potential susceptibility to subsequent closure by a change of radwaste disposal policy, namely, a shift to surface storage instead of disposal. NIREX did not seem to possess this long term perspective in the mid-1980s. Anyway, the plans to bury short-lived intermediate and low level wastes at the Elstow Storage Depot and at the Bradwell, Fulbeck and Killingholme sites aroused intense local opposition even after intermediate level wastes were excluded from the NIREX proposals. Opposition appeared on two fronts: the local public interest groups under a national 'parent' organisation, BOND (Britain Opposed to Nuclear Dumping); and political interest groups comprising local MPs and the affected local authorities. Given the government's tacit support for NIREX, it was clear that this opposition could not win in the short term but it did demonstrate the futility of trying to push through what seem to be massively publicly unpopular and unacceptable siting proposals even when the investigations were at an early stage. Such public reaction is also typical in other countries, particularly the United States once potential siting areas have been identified.

The original announcement of NIREX's intentions to investigate sites at

Elstow in Bedfordshire and Billingham in Cleveland in October 1983 had resulted in the local people forming themselves into opposition groups to present a united front against the proposals. Both groups took the same acronym, BAND (Bedfordshire/Billingham Against Nuclear Dumping) and developed close links in their 'fight' against NIREX. In late 1985 when its became apparent that the sites at Bradwell, Fulbeck and Killingholme were to be included as additional sites for study and comparison along with Elstow, three more local opposition groups were born. These were EAND (Essex Against Nuclear Dumping), LAND (Lincolnshire Against Nuclear Dumping) and HAND (Humberside Against Nuclear Dumping). When these three additional sites were officially announced on 25 February 1986, EAND, LAND and HAND together with BAND formed themselves together under the single 'umbrella' organization of BOND (Britain Opposed to Nuclear Dumping). The exclusion of ILW had had no significant effect.

The BOND network of anti-dumping groups was born out of the National Conference on Radioactive Waste Dumping held in Bedford in April 1986 and was officially launched at a press conference held in the House of Commons on 22 October 1986. The network, in their own words, 'represents unity and solidarity [between organisations], and shows that the campaigning groups are deeply concerned about the whole principle of the shallow land burial of low level waste, rather than being caught up in any 'Not In My Back Yard' (NIMBY) argument' (BOND, 1987). The main aims of BOND were to secure an immediate halt to all site investigations; to advocate interim above ground storage of nuclear waste at its place of origin; to demand that the government implement a safe, well researched and publicly acceptable policy on nuclear waste; to alert the public to the dangers of low level ionizing radiation and to require complete openness and public accountability in all matters relating to nuclear waste disposal. Only the first of these aims has been realized in full, while the health risks from low level radiation are still not fully understood. These BOND suggestions are really quite sensible The idea of surface storage is important because of the implicit risk that a change of government would readily adopt it as an easy solution to the problem, although it might be imagined that surface storage is still likely to be unattractive to communities located nearby. More important, however, would be the prospect of locating a surface waste storage facility virtually anywhere in Britain, without too much regard for geology. This should presumably make it easier to find a publicly acceptable site.

## Local authority opposition

The local authorities concerned were also quick to react. Three of those involved, Bedfordshire, Humberside and Lincolnshire County Councils, entered into a consortium, the County Councils Coalition (CCC), to present a joint front of opposition against NIREX's proposals. This Coalition was potentially a far greater risk to NIREX. Previously local authorities had tended to take a passive role in nuclear planning, relying heavily on external expert advice provided by governmental agencies. Now there seemed to be political support for a much more aggressive approach. This was a problem for NIREX because of the limited resources available to them reflecting their function as an 'executive', although of course they could presumably call on the much greater resources of the UKAEA and CEGB. Nevertheless, strong, determined, local authority based opposition with strong public support does provide a far more able adversary than fringe-minority based anti-nuclear groups.

The CCC found the proposals of NIREX unacceptable for the following reasons:

(1) In their view shallow land burial (SLB) had not been established as the BPEO for LLW and no satisfactory evaluation of alternative methods of disposal had been carried out;
(2) NIREX's plans were poorly thought out and contained many scientific uncertainties regarding the long term safety of the SLB concept;
(3) the existing waste classification is unsatisfactory and it was not clear which wastes were intended for disposal by SLB;
(4) there are no national agreed site selection criteria for radioactive waste disposal facilities; and
(5) insufficient regard was paid to public opinion by both NIREX and the government.

In this wider context of the national waste management policy the CCC picked up on the views and findings in the report of the all-party House of Commons Environment Committee (1986).

The CCC believed that for a particular disposal site to be accepted it must be part of a coherent and publicly acceptable national waste management policy. Although the ILW component was dropped from the proposals by the government in response to the criticism in the Environment Committee's report, the CCC maintained that the government's answer contained in its White Paper *Radioactive Waste* (DOE, 1986a) was still inadequate and the dropping of the ILW element only served to demonstrate the *ad hoc* nature of their decision making.

The Environment Committee did conclude that whereas the poor state of research in the United Kingdom made it impossible to recommend any disposal option with full confidence, the near-surface option would be acceptable if fully engineered to ensure complete containment, but only for short-lived LLW. Additionally, much greater emphasis should be placed on research and development into sub-seabed options for ILW similar to those being considered at present. There is a very important point here. There is a major difference between knowing how to handle the radwaste problem in a general conceptual fashion and knowing how to operationalize it in reality. There was a gradual realisation, although not yet among all MPs, that NIREX would grope their way to an operational solution while actually building radwaste dump No.1. There is no prototype and the first prototype has to be the production version. The apparent paradox concerns the realization that while there are no technical problems in disposing of waste this is only true in a theoretical sense and there may well be horrendous shocks and surprises in engineering a real-world facility that has to work and meet the high standards of safety that have been set. Since no other country seems to have successfully built such a facility so the NIREX optimism is probably somewhat misplaced. On the other hand, what else could they possibly say at this stage of the game?

The CCC also drew comparisons between the deep disposal facilities under construction in Sweden and West Germany for LLW and ILW and those proposed by NIREX in the United Kingdom. The CCC, like the Environment Committee, expressed concern over the apparent lack of consideration of the deep disposal option. Regarding the more specific question of the shallow land burial (SLB) disposal scheme developed by NIREX, the CCC dismissed it as being totally inadequate, simplistic and fundamentally flawed since operational experience in the United States has demonstrated that SLB is only a viable option in arid environments (unlike our wet climate). The CCC was also critical of the official waste classification system and its application. They regarded the range of wastes included within the categories of low and intermediate as being too high since they only take the level of their activity into account and not the type of radiation or its longevity and biochemical toxicity. The CCC also maintained that a precise inventory should be made of the wastes to be disposed of at a repository before its development and not just a broadly defined waste stream such as 'LLW' as put forward by NIREX, otherwise there can be no hope of establishing public confidence in the operation.

On the subject of site selection procedures the CCC were again very critical. It is plain to see that all the near surface sites identified by NIREX are in public ownership (three owned by the CEGB and the fourth by the MOD). This suggests that the site selection process had been biased by the dominant factor of site ownership which are not particularly important criteria from a public safety perspective. Here again the CCC is in support of the Environment Committee in calling for advanced publication of site selection procedures and criteria so as to remove any possiblity of a 'fudge factor' in siting decisions. The CCC go further saying that: 'If any community is to be expected to host a radioactive waste repository on behalf of the nation, then it must be shown in advance that the location has been arrived at through a nationally agreed, logical process with public safety the prime consideration.' The way forward, according to the CCC, is via the disposal of all our radioactive wastes in deep repositories and by reducing the amount of LLW arisings by sorting, incineration and compaction. Until a suitable location for a deep repository is found it should be possible to continue using the Drigg site, which has capacity for at least another 20 years. The CCC maintained at the time that the additional costs incurred from routing all wastes to a deep repository appeared insignificant and would be a small price to pay for public acceptance.

Subsequent events may well have proven them right.

In November 1986 representatives from the CCC were among an 18 strong team which visited radioactive waste repositories in West Germany, France and Sweden. A report, based on the findings of the tour, was published in January 1987 by Environmental Resources Ltd. It found the proposed shallow land burial (SLB) concept of waste disposal was ill-conceived, inadequate and potentially unsafe compared with the deep burial methods employed by Sweden and West Germany. The visitors were most impressed with the facilities at Forsmark in Sweden where low and intermediate level wastes are to be disposed of hundreds of feet below the seabed off the Baltic coast. The team was less impressed with the facilities at Cap de la Hague in France, the example of a shallow burial facility much praised by NIREX. Here the team says they found sloppy safety procedures and waste disposed of in damaged containers (The *Guardian*, 23.1.87). The three primary findings of the tour were summarised in the report as follows:

(1) The United Kingdom is well behind in the development of facilities for radioactive waste disposal in comparison to our European neighbours;
(2) the relative costs of the different methods of disposal are more similar than suggested in the DOE's BPEO report, thereby invalidating SLB as the significantly cheaper option; and
(3) the openness of the disposal organizations visited contrasted sharply with the experiences of the CCC members in attempting to deal with NIREX.

Finally, after attacking the general principles underlying NIREX's operations, the local authorities also made their own site specific cases against the development of a SLB facility for LLW in their areas.

There is little doubt that if the objective is to oppose a radwaste dump then most of the effort has to be directed against the early investigative stage when there are alternative locations and little site specific investment by the developer. Once the public inquiry phase is reached then the chances of success are at best minimal. However, forming a consortium dedicated to blanket opposition is not likely to be successful since it directly conflicts with the national interest arguments that NIREX use to support the need for a single national site. The government would simply support NIREX and favour whatever location they wished to develop. If there is in fact a choice between sites, then NIREX would no doubt be influenced by the intensity of local opposition. From the point of view of an individual local authority, the best selfish strategy would have been to muster as much explicit documentary evidence of local opposition as possible; for instance, holding referenda on the NIREX proposals. Another possibility, would be deliberately to seek to deny potential sites by proposing developments which are explicitly and deliberately incompatible with a radwaste dump. For example, ground water extraction, new housing developments, a change in landscape designation, a proposal for an oil storage depot or a hospital, and even a major retail development which would create traffic congestion and thereby 'manufacture' a possibly unacceptable site.

## Public direct action

Attempts by NIREX contractors to start site investigations at Bradwell, Elstow, Fulbeck and Killingholme began in mid-August 1986 soon after the Special Development Orders (SDO) enabling NIREX to proceed had been passed by Parliament on 14 May of that year. It was not until mid-September 1986 that drilling contractors managed to gain access to all the sites and start work. In the months between July and September local villagers had maintained a vigil, sometimes working in shifts around the clock, to stop NIREX's field investigations going ahead as planned. Villagers at the four sites were warned by NIREX a week before investigations were due to start that protesters attempting to prevent exploratory drilling going ahead would face court injunctions. However, as contractors moved in to start drilling work at Elstow, Fulbeck and Killingholme on 18 August hundreds of local protesters blockaded site entrances by sitting cross-legged in the road. By the end of the month villagers were still successful in blocking the sites to NIREX contractors.

On 2 September 1986 NIREX lost patience and announced that time had run out for the protesters and that legal action, only a possibility a month ago, was now 'seriously under consideration' (*Daily Telegraph*, 3.9.86). This warning came on the same day that contractors got onto the Elstow site through the unwatched entrance. The drilling equipment stayed on site guarded by security men until operations were completed. At Bradwell, where investigations were due to start on 1 September, villagers and their children had blocked the road preventing contractors from reaching the site. The blockading of the Bradwell site lasted until 16 September when police cleared the way for contractors to drive their vehicles and equipment onto the site, foiling the vigil of seven locals and their attempts to call for reinforcements. Meanwhile NIREX contractors had gained access to the sites at Fulbeck and Killingholme. With the contractors getting onto the four sites a few protesters changed their tactics. Six protesters halted drilling work at the Killingholme site on 21 September by breaking into the contractors compound and staging a sit-in on top of a freight container. The demonstrators were removed by the police after 23 hours. Other protesters restricted their actions to lawful protest outside the sites, such as silent vigils and the scattering of flowers in the wake of contractors' trucks.

In the end access to all four sites was gained peacefully with no injuries and no arrests, although court injunctions were issued against HAND and LAND. The public opposition demonstrates another paradox. People are perfectly entitled to protest but NIREX are also entitled to perform their site investigations, the difference being that peaceful protests are essentially unlawful no matter what degree of provocation, while the NIREX operations are protected by the law. The implications are quite clear. People interested in protesting against NIREX operations need to do battle elsewhere than on the site itself. The latter can only be a short-term delaying tactic, in the end it will fail and history shows that it has previously always failed even if it makes people feel better. NIREX in fact displayed commendable patience but they still need to be aware that their 'legal backing' is dependent on the support of government and it is not a divine right they possess. Each confrontation serves mainly to make the NIREX task more difficult, increases the level of politicalization, and serves to distract from the basic national needs and scientific arguments. There are limits as to how far publicly unpopular developments can be pursued in Britain. It is one thing to have a small vocal minority opposed to a modern development, it is quite another when the small fringe minority becomes a large popular minority and even a majority. Nuclear power developments never previously aroused so much opposition as radwaste dumps.

### Round Three: *The Way Forward* (1987)

On 30 April 1987 John Baker, chairman of NIREX, wrote to Nicholas Ridley, Secretary of State for the Environment, recommending a major change of approach towards the disposal of LLW. In his letter Mr Baker drew attention to the increasing cost estimates for the shallow land burial (SLB) of LLW and how the government's decision to exclude short-lived ILW from SLB had increased the unit cost of disposal by this method. Mr Baker concluded that there would be little difference in cost if LLW were to be 'piggy-backed' into a deep repository along with ILW. NIREX therefore recommended that site investigation work be abandoned at the four candidate SLB sites and effort concentrated on identifying a site for a deep repository.

Findings from site investigations had revealed that the geological sequences at the four sites were basically as predicted. Water movements through the clays at each site are extremely low, but the hydrogeology at all the sites, especially Killingholme, was more complex than previously thought. At Killingholme a metre-thick permeable sand layer runs across the site eight metres below the surface giving rise to less than desirable groundwater flows. Should a repository have been built at Killingholme it would have required 'additional expenditure on the "near-field" barriers to give equivalence with the other sites with lower groundwater flows' (Baker, 30.4.87). One area of uncertainty relates to gas generation. Studies have shown that over long timescales the production of hydrogen from corroding steel and carbon dioxide and methane from the decomposition of organic material such as paper and wood could be quite large. In an impermeable clay site this could cause a significant build up of pressure, although how this occurs and over what timescale is not known.

It is the view of NIREX however, that a repository which meets the safety criteria laid down by the DOE could be engineered at any of the four sites. The main point of the letter however, was that costs had risen due to the exclusion of short-lived ILW, the extension of site surveys and pressure to provide a 'Rolls Royce' type of SLB. These developments had been driven by the need to respond to public perception rather than any technical requirements. At the time of the letter NIREX's best estimate of the SLB was £500 – £1,000 per cubic metre of raw waste whereas the estimated cost of disposing of ILW in a deep repository stood at about £2,500 – £7,000 per cubic metre depending on the type of repository used. By combining LLW and ILW disposal into one deep facility the cost of disposal for LLW is estimated at £750 – £1,200 per cubic metre, not much different to estimates for SLB and certainly less than the reliablity of the figures themselves. It was therefore the suggestion of NIREX that all the LLW and ILW should be disposed of together in one deep repository either at a land-based site, in offshore tunnels or beneath the seabed using offshore technology.

The SLB option does not however seem to have been laid totally to rest; it may well be revived later for decommissioning wastes. These large items, weighing many tonnes, are only lightly contaminated and their activity is very short lived. This suggests that the additional costs of cutting them up for deep disposal are probably not justifiable. The other favoured disposal route for bulk decommissioning items mentioned in the BPEO report is deep sea disposal. Both SLB and sea disposal have been shown to be unacceptable in the public eye, but may conceivably be resurrected in the future. The implications are that one, maybe more, of the four land sites could be reconsidered at some future date. Their local communities may wish to consider what steps they can take during the next five to ten years to preclude this possibility. It should be remembered that the site investigations were completed and doubtlessly the results will not be wasted.

The day following Baker's letter, just 10 days before the dissolution of Parliament in the run up to the 1987 general election and without consulting RWMAC, Ridley suprised MPs in the House of Commons by announcing that he had accepted Baker's recommendations. Opposition MPs were sceptical and although welcoming the decision were quick to criticize the timing of this apparent U-turn in waste management policy. Opinion polls in the four constituencies affected had apparently shown wavering support for the Conservatives in areas where they normally held a strong majority. Labour's environment spokesman insisted that the decision was made by the government 'in a squalid attempt to save themselves from electoral embarrassment' (The *Guardian*, 2.5.87). In fact, if NIREX had wanted to continue with one or more of the sites they would have received government support. However, once they had decided to abandon the concept of a LLW facility then there was no advantage in delaying this decision until after the General Election of 1987.

At last it seemed that NIREX had grasped the basic need for public acceptance as a basic criteria of equal importance to more purely technical aspects. As has been commented upon in previous sections, tacit public acceptance both now and in the future are essential pre-requisites for a successful national radwaste dump. Steam-rollering unpopular sites through the decision making process, while feasible, will not provide a stable foundation for the long-term future. All party support is essential, at least for the next 30 – 50 years, after which maybe NIREX will be too large and too well established to be so readily manipulated by political factors. In *The Way Forward* (NIREX, 1987) document, NIREX seem to have replaced or modified the traditional deeply entrenched nuclear industry arrogance with a new pragmatism that has a good chance of being successful. In particular, the new NIREX asked for advice and comments about how it should proceed within the framework of government policy.

The current disposal policy is that all low and intermediate level wastes should now be placed together in a deep mined repository, either on land, under the seabed with tunnel access from a coastal drift or under the seabed from an offshore rig. Compared with the earlier suggestions, these are 'Roll Royce' options which represent improved safety and with it , potentially greater public and political acceptability. There will no doubt, however, be an element of NIMBY when sites are again short-listed. Even with an offshore rig-based facility waste traffic will have to converge on one or two ports in somebody's back-yard to be shipped out and even then the seas are, to quote Holliday (1984), 'the fishermen's back-yard'.

To increase local acceptance, and this NIREX now admit is important, it is also necessary to increase local involvement in the decision making process of policy formulation and site selection. It now appears that the government and NIREX are beginning to recognize the importance of public acceptability. In his letter to Ridley, Baker wrote: 'it would be wise to avoid premature commitments as to a preferred technique or how and when sites for investigation are to be identified. One lesson from our work so far is surely that the public does not like feeling pressurised to accept imposed solutions in this area and … time spent now in considering all the issues in this way would be time well spent.'

Indeed NIREX have now gone some way towards achieving their aim of a more open site selection strategy. In their discussion document, *The Way Forward*, NIREX point to the need to develop a new deep repository, the large sums of public money involved and the impact such a facility would have on the local community as creating the need for open discussion and feedback on all the related issues of siting, design and environmental protection from the British public. In tendering people's opinions NIREX asked them to address a number of key issues. Which of the three disposal options commands support? Which factors should be taken into account when selecting a site and how should they be weighted? What weight should be given to the views of neighbouring countries when considering sub-seabed options? Should important 'amenity' areas, such as National Parks, be eliminated from the very start of the site search? What locations should be subject to detailed consideration? How important is the retrievability of disposed wastes? In addition NIREX invited suggestions on how it could involve itself in the local community, how local interests could best be represented and requested ways of helping a NIREX development become a good neighbour and so smooth out NIMBY type problems. The discussion document also invited interested organizations to meet NIREX representatives for extended discussions. Some 50,000 copies of *The Way Forward* were distributed between November 1987 and June 1988 and a total of 2,526 replies were received.

In their search for the best location for a deep repository NIREX have been following the basic guidelines laid down by the IAEA (1983). The IAEA suggest a three stage site search comprising a national survey, preliminary site identification and site confirmation. Although public discussion of the issues at hand is at present underway, it is not clear from the document how and when public participation and consultation will come into this site selection process. The best NIREX can hope for is that some communities will actively bid for a radwaste dump either as a means of rejuvenating local economic prospects or because there is some prospect of obtaining some local employment benefit. A large underground storage facility would be a major engineering undertaking. It would have a long operational life and might well be considered a far more useful local asset than a near surface dump. The most favoured sites are at Sellafield and Dounreay. In both cases there is likely to be local support, rather more perhaps at Dounreay where the possible closure of the FBR will create a major local unemployment problem. These locations are very sensible because it is here where most of the wastes are generated and they are also prime locations for the future development of additional reprocessing capacity.

### The way forward or the way back?

NIREX started out with what was essentially an impossible task. They have done extremely well to reach the position they have. The task is not technically difficult and there are several acceptable solutions to the radwaste disposal problem. However, the problem is politically extremely difficult – maybe even impossible – because rather than being a purely technical matter it is one of public acceptability in which emotion and irrational fears have replaced scientific logic. The government in creating NIREX was probably more aware of these difficulties than the industry was at that time. They were also clever enough to see the need for a tool such as NIREX which could be used to distract potential criticism from themselves in facing up to a problem that wins no votes. No one wants to live near a conventional rubbish dump let alone a radioactive one!

The three siting rounds to date reflect a gradual realization of the realities of the public acceptability problems. The first site proposals were a minimum cost strategy by a young NIREX full of enthusiasm for the task. The second attempt was more expensive but it was still an engineer's solution to a problem that by then was certainly no longer a purely technical affair. The third attempt represents a potentially much more expensive but also more broadly balanced approach. Clearly NIREX are following a learning curve and they have made some mistakes. But there are now signs of progress and they do seem to have moved on from a technocratic view of the world to a more flexible and pragmatic approach based on the political realities of late-20th-century Britain. No doubt there are still important lessons to be learnt. For instance, there is still a reluctance to avoid what is regarded as the 'premature exclusion of areas or regions from possible use ... since that in turn only encourages a competition to name areas of search before the necessary work has been done' (Baker, 1987) – for no good reason. There is little doubt that large parts of Britain are potentially unsuitable on either technical or public acceptability grounds. It would do little damage to any conceivable NIREX scheme to declare, for instance, that sites under urban areas are excluded. Nevertheless, it is good that an element of public consultation has now been introduced but it remains to be seen to what extent public participation can be implemented into the site selection process itself. It would still be easy to revert to a revised round two approach bolstered up by claims that there was now public support for whatever proposals happened to be made. It would be nice to think that this will not happen because it will not work.

There is also the risk that NIREX might become frustrated with the problems of seeking a publicly acceptable solution and revert to the traditional nuclear engineer's steamrollering solution. This attitude of mind certainly still exists. For example, at a radioactive waste management conference held in 1988, it was fairly obvious that there is a sense of outrage within the nuclear industry at the measures being taken to appease public opinion; presumably this is seen as unnecessary as public opposition has traditionally been ignored by the nuclear industry. The former head of the National Radiological Protection Board is quoted as commenting that the industry has been stupid enough to give in step-by-step so that it will now have to provide an expensive Rolls-Royce solution: 'The proper engineering solution should be put forward ... if anyone doesn't like it they should say what they want and then pay for it' (Quick, 1988, p.21). Whether NIREX can withstand such extreme arguments by invisible but industry influential people and achieve a broadly based commonsense solution is still in doubt.

# 5

# Britain's radwaste dumps: past, present and proposed

This chapter is concerned with providing some details of where Britain's radwaste has and is being dumped. The two principal disposal routes have been the northeast Atlantic and Drigg. In addition, details are given of the characteristics of the various sites that have been proposed at Billingham, Elstow, Fulbeck, Killingholme and Bradwell. The latter are still important not so much now but in the longer term. It has been stated previously that it should not be assumed that one site will meet all Britain's radwaste dumping needs. Obviously NIREX are now firmly focused on securing the development of only one facility, but once permission has been granted then there will be pressure to seek additional sites to handle bulky decommissioning wastes. There are also other factors that may intervene; for example, clean up after a reactor accident in the United Kingdom would require urgent access to near surface and deep dumping facilities. The MOD also has its own requirements and may well operate in parallel with NIREX. It has been noted that the 'play fair' constraints that impinge on NIREX are absent from the MOD. The MOD can more or less do whatever it wishes without worrying about land ownership and planning constraints. If NIREX do not wish to use some of the round two sites then the MOD may well do so. So to some extent the existing proposed sites have been reprieved in the short term but their future remission is still in doubt. They are, after all, the best understood locations in Britain. It is worthwhile therefore in describing in some detail their principal features, noting where relevant any additional factors that may be involved. First though it is useful to examine the full range of disposal options.

## Disposal options

Broadly speaking there are five principal waste disposal options: sea disposal in the northeast Atlantic dumping grounds, two types of near surface disposal (shallow land burial and engineered trench disposal), deep cavity disposal under land or under the seabed with access from land, and disposal via off-shore boreholes into the seabed. The possibility of long term or indefinite storage on land is not considered to be a sensible option. Sea disposal is the simplest approach to the problem. However, it is fairly expensive due to packaging and transport costs and only a proportion of all ILW are suitable (about 20 per cent). Nevertheless, it is claimed to be a safe, well understood option that presents minimal long term risks. The wastes are not considered to be retrievable. Containment is by dilution and inaccessibility. The wastes are merely dumped on the deep seabed in the hope that they stay there and only return into food chains and the terrestrial environment after massive dilution. Shallow land burial (SLB) is personified by Drigg. The waste is buried in a shallow trench under at least one metre of top cover within a clay environment. There is no packaging or compaction. Containment depends on the properties of the clay and a balance between leachates and half-lives. It is no longer an acceptable practice for a wet climate but it appears to have worked well at Drigg, so far. A more sophisticated variant of SLB is engineered trench disposal (ETD). This is considered suitable for all kinds of LLW and short-lived ILW which have been conditioned by immobilization in cement. The latest vault developments at Drigg might be considered to be a variant of this approach. The trench could be in clay or some other geology and might well be cement lined. Monitoring and surveillance would continue for 300 years.

The longer lived ILW require a mined cavity at 100 – 300 metres' depth to ensure isolation from the environment for the longer timescales that are now necessary (Griffin *et al*, 1982). This deep cavity disposal (DCD) facility might be excavated under land or under the sea with access via a tunnel from a land base. As with the other forms of land disposal, recovery of the wastes might just be technically feasible but expensive, time consuming and difficult. The assumption is made that recovery would not be attempted. The depth of DCD would minimize risks of accidental disruption and would provide additional barriers to escape. However, the packaging would not survive long enough to contain the radionuclides and containment efficiency would depend on the length of migration route back to the surface. In practice this should not be a problem; except perhaps on very extended timescales (say 50,000 years or so). It does, however, raise the question as to how best to mark these disposal locations for future generations.

Disposal in off-shore boreholes beneath the seabed is another possibility, if, the potential legal problems can be solved; for example, who owns the seabed? A metre depth hole to a depth of 2,000 – 3,000 metres would be filled with ILW and then plugged with cement. The wastes would be non-retrievable. A final possibility is that of long-term storage prior to disposal. This offers a potentially useful public-anxiety-reducing option. Ultimate disposal could be delayed for up to perhaps 200 years. However, the storage period has to be credible in terms of the monitoring, maintenance, and surveillance commitments it would entail. Do we have the right to leave behind such legacies that may well require active attention? This option is currently out of favour because it entails ongoing maintenance costs, it presents greater theoretical risks to the workers, and above all it is considered unnecessary while disposal options are considered to be feasible.

*Table 5.1* Times at which risks are highest for various disposal options

| Disposal method | Maximum risk period |
| --- | --- |
| Sea Dumping | Environmental dispersion 100 – 1,000 years |
| SLB and ETD | Intrusion or disruption after 300 – 1,000 years<br>Dispersion after 500 – 1,000 years |
| DCD | Intrusion after 300 – 1,000 years<br>Dispersion after 25,000 years |
| Off shore borehole | Intrusion after 300 – 1,000 years<br>Dispersion after 12,000 years |

Source: DOE (1988b) p.21.

Table 5.1 gives estimates of the times at which maximum risks might occur for each of the disposal options. The greater intrinsic safety of DCD should be noted. The co-dumping of ILW and LLW via DCD would seem to offer the maximum possible levels of containment.

## Dumping radwastes in the deep ocean

Sea disposal is by far the easiest and cheapest option. It is hardly surprising therefore that the disposal of packaged solid radioactive wastes into the northeast Atlantic began in 1949 and continued until 1983. Disposal operations were the responsibility of the UKAEA up to 1983, when it was handed over to NIREX. The seventh meeting of the London Dumping Convention also took place in 1983 and although a proposal to prohibit the deep disposal of all radioactive wastes was unsuccessful, it gained the backing of the National Union of Seamen and other transport unions. The boycott which followed prevented the planned 1983 operation to dispose of 3,500 tonnes of packaged wastes using the specially converted ship, the *Atlantic Fisher*, from taking place and has effectively put a stop to the United Kingdom's annual sea dump ever since. Whether this ban continues indefinately is important because it critically affects the amount of space needed for land based disposal in the future. Indeed, in 1988 the British government finally agreed to abandon sea disposal as a route for ILW, but wished to retain this option for the bulkier items of decommissioning wastes.

Between 1949 and 1982 the United Kingdom admits to having disposed of approximately 74,000 tonnes of packaged low and intermediate level waste (totalling approximately 721,000 Ci or 27,000 TBq of radioactivity). Most wastes were packed in oil drums filled with cement and dropped over the side of a ship. These packages were only designed to remain intact during descent to the seabed and were thought likely to release their contents slowly over an unknown period as part of the 'dilute and disperse' principle of sea disposal.

Solid low and intermediate level wastes have been dumped at nine sites in the northeast Atlantic. These sites are more than 4000 metres deep, they are free from up-welling currents and they are remote from fishing grounds and submarine cables. The first operation in 1949 was carried out on an experimental basis with subsequent disposals approximately every two years up to 1961 when they became an annual event. From 1967 to 1976 disposal at sea became a joint operation, coordinated by the Nuclear Energy Agency (NEA) (then the European Nuclear Energy Agency, ENEA) of the OECD. This organization involves Belgium, France, the Netherlands, West Germany, Italy, Sweden, Switzerland and the United Kingdom. Only the United Kingdom and the Netherlands took part in every operation, while the United Kingdom also carried out separate operations in 1968 and 1970. Since 1977 and up to 1982, there has usually been two annual dumping operations in the northeast Atlantic, one by the United Kingdom and one jointly by Belgium, the Netherlands and Switzerland. These were run independently from NEA, but in conformity with the London Dumping Convention (LDC) and OECD/NEA regulations. The present site, which was selected by NEA in 1971 to meet the disposal needs of the various OECD member states, was used exclusively between 1971 and 1982. The site comprises a 4,000 sq km rectangle, 700 km of the west coast of Spain and 950 km southwest of Lands End. The average depth is 4400 metres or 2.5 miles. All dumping stopped in 1983.

One immediate problem arising from the abandonment of the British sea dump was the drums of radioactive waste originally intended for disposal in the 1983 operation which are currently stored at Harwell. These waste packages, while suitable for sea disposal, would probably not come up to the standards required for deep disposal on land. While the UKAEA maintains that repackaging is not feasible, it may be possible to 'overpack' the Harwell drums at some future date. Meanwhile, other waste that would have been dumped at sea has been diverted to Drigg.

Authorization for sea dumping is given by MAFF and the DOE under the Radioactive Substances Act (1960). This includes both liquid and gaseous discharges and solid disposal. Disposal through the National Disposal Service (NDS) of radioactive wastes from non-nuclear licenced sites requires authorization from the DOE only. Authorization covers the amount of waste to be dumped, both in weight and activity, waste form, the disposal method and the timing of disposal operations. It is purely an administrative device rather than a safety measure. The United Kingdom is also signatory to two major controlling agreements governing the disposal of radioactive wastes at sea.

These are: the 1972 Convention on the Prevention of Marine Pollution by Dumping of Wastes or Other Matter (more usually known as the London Dumping Convention or the LDC); and the 1977 OECD/NEA Multilateral Consultation and Surveillance Mechanism for the Sea Dumping of Radioactive Waste (known as 'The Mechanism').

The provisions of the LDC currently control the disposal of all wastes by dumping into the oceans and came into force in 1975 with about 50 state signatories, including the United Kingdom. The LDC aims to prohibit the disposal of certain kinds of waste and allow the disposal of others only after a special permit has been issued by the competent national authority responsible. Through the LDC the deep sea disposal of high level wastes, as defined by the IAEA, has been declared unsuitable and is banned, although that does not mean that there has never been any or that Britain still retains in storage its total stocks of high level wastes.

The requirements of the LDC were given legal status within the United Kingdom by the Dumping at Sea Act (1974), whereby any ship sailing from an English port has to have a licence issued by MAFF. Ships sailing from Scottish or Welsh ports need licences issued by the Department of Agriculture and Fisheries for Scotland or the Welsh Office, respectively. Again this applies to both nuclear and non-nuclear wastes. Any dumping permits granted have to be declared to the LDC through its secretariat, the International Maritime Organization (IMO). The issue of licences must take into account all International Atomic Energy Agency (IAEA) recommendations regarding the sea disposal of radioactive wastes, including those describing limits on the tonnage and activity of wastes to be dumped, dump site selection, dump ship facilities, waste packaging, supervision of operations by escorting officers, record keeping, international cooperation, monitoring and environmental assessment. Recommendations by the IAEA describing the limits imposed on the activity and tonnage of wastes dumped at sea are perhaps the most important. These have been reviewed on a number of occasions. The first major review concluded that low and intermediate level waste could be safely disposed of at designated sites at least 2000 metres deep. Current limits put forward by the IAEA under the terms of the LDC specify sites more than 4000 metres deep and are made more complex by reference to radioactivity per tonne of waste and set limits on the types of radioactivity and on specific radionuclides.

The OECD Mechanism has currently been signed by over 20 member states and was established by the OECD to further the objectives of the LDC by providing a route for consultation between member states, via NEA, prior to dumping operations. The Mechanism states that NEA must be informed, at least six months in advance of dumping, by the national authority responsible. In the case of the United Kingdom, this is done via the DOE. In addition to administering this prior consultation exercise, the Mechanism also provides for: the establishment of guidelines and recommended practices and procedures for the safe dumping of radioactive wastes; international observation of dumping operations by an NEA representative; and a review of the suitability of disposal sites at least once every five years, in other words a gentleman's agreement about the rules of the sea dumping game. The entire process was assumed to be safe until there was evidence to the contrary.

At the meeting of the LDC in February 1983, a Spanish resolution was passed calling for a voluntary suspension of all sea dumping of radioactive waste pending the outcome of a number of international scientific reviews. Although the resolution was not legally binding NIREX was prevented from carrying out its planned disposal operation through protest action by transport unions, who supported the voluntary ban. Prior to 1983 the environmental organization Greenpeace had been campaigning against the dumping of radioactive waste in the northeast Atlantic since the mid 1970s. Each year Greenpeace had lobbied the annual meeting of the LDC to focus attention on their opposition to the continued dumping of nuclear waste. In 1982 they made a well publicized attempt at direct intervention by using inflatable boats to try and stop dumping from taking place.

Despite obvious opposition, the British representatives to the LDC voted against the cessation of the annual sea dump because the resolution was against government policy and 'against the spirit of the convention' by calling for a halt in operations before receiving relevant scientific advice (Alec Buchanan-Smith, Agriculture Minister, 22 February 1983). For Greenpeace and other opposing groups, however, it was not so much the pure scientific advice as the administrative philosophy of the nuclear waste dumping strategy that was the highest concern. That philosophy of a projected increase in the use of sea disposal, backed up by appropriate new scientific evidence to provide the necessary international acceptance, was outlined in the 1979 report *A Review of CMND 884*. This was seen by Greenpeace as an attempt to pervert scientific analysis to provide justification for the continued use by the United Kingdom of the sea disposal route (Blowers and Lowrey, 1985).

In June 1983 the three leading transport unions, TGWU, ASLEF and NUS announced that they would boycott the 1983 sea dump planned for 11 July. An appeal to the unions by the owners of the dump ship *Atlantic Fisher* failed to alter their decision. Even in the light of all the opposition, both from Greenpeace and the unions, the British government did not abandon hope of continuing with the the 1983 dump and still argued for its revival. By August it was reported that the government had abandoned all further plans for dumping nuclear waste at sea which was hailed as 'a remarkable victory for Greenpeace' (The *Observer*, 28 August 1983). When, in September 1983, the leader of the NUS, Jim Slater, won over a resolution to boycott all future sea dumping of radioactive waste at the annual TUC conference, the option to re-open dumping plans was effectively removed from government waste management strategy.

In December 1983, a joint Government–TUC committee was formed under the chairmanship of Professor Fred Holliday to investigate sea dumping and its potential dangers. When the Holliday Committee reported a year later it recommended the LDC moratorium on sea dumping remain in force until the completion of three ongoing international reviews by NEA, IAEA and the LDC (Holliday, 1984). These were:

(1) the OECD/NEA five-yearly review of the suitability of the northeast Atlantic site;
(2) the review of the IAEA definition for the LDC of high level wastes unsuitable for sea disposal, and the recommendations to national competent authorities about permitted disposals; and
(3) the *ad hoc* review of the scientific and technical considerations relevant to sea disposal being carried out for the contracting parties to the LDC.

The Holliday Report also recommended that a comparative assessment should be made and published of all disposal and storage options, whether sea or land, in order to establish the best practicable environmental options (BPEOs) for various types of waste. It did not make any suggestions as to the permanent status of sea disposal despite having found no evidence to show that sea disposal is hazardous.

By the end of 1985 all three international reviews awaited by the LDC had been completed. The NEA five year site suitability review was completed in April 1985, concluding that: 'even if the practice [sea disposal] were going to be continued for 500 years at present rates, doses to individuals would still be more than three orders of magnitude below the appropriate limits', and that: 'On the basis of the evidence presented in this review, the site could be used for dumping of packaged, radioactive wastes during the next five years', until the next NEA site review (OECD/NEA, 1985).

The IAEA review of the definition and recommendations for the LDC reported in September 1985. The revised definitions and recommendations contained nothing which would preclude the resumption of sea dumping of permitted wastes by the United Kingdom. The *ad hoc* scientific review for the LDC undertaken by an international panel of scientific experts drawn from the International Council of Scientific Unions (ICSU) and the IAEA, had reported in draft by early 1985. An expanded panel revising the draft were unable, however, to agree on the conclusions to be drawn from it, although the majority considered that there was no scientific reason for the LDC to prohibit the disposal of radioactive wastes into the sea.

At the ninth meeting of the LDC, following the completion of these reports, most member states declared themselves to be against the dumping of radioactive wastes at sea in principle, regardless of the scientific evidence, and wanted the voluntary ban to continue pending yet more complex studies. The British delegation made it clear that the government believed, on the basis of all the scientific evidence to date, that sea disposal would continue to be one of a range of available options for low and intermediate level radioactive wastes, but was committed to await the outcome of the BPEO study called for in the Holliday Report before coming to a conclusion on sea disposal. From this it appears that the government was still holding out hope for the eventual resurrection of its annual sea dump even after announcing, two years before, that all further plans had been abandoned.

When the BPEO Report was published by the DOE in March 1986, it expressed a broad support for the government's waste management strategy including sea disposal for MAGNOX ion-exchange resins, plutonium contaminated materials (PCMs), tritiated wastes from commercial and medical applications of radioisotopes and some decommissioning wastes (DOE, 1986b). After reviewing all the above evidence, RWMAC came down strongly in favour of sea disposal for selected wastes and recommended that the 1983 consignment be disposed of at sea as soon as possible (RWMAC, 1986).

Even though numerous scientific and technical studies have declared that the sea disposal of selected radioactive wastes is a safe and viable option, the majority of signatories to the LDC and OECD/NEA Mechanism are opposed to its use in principle and are in favour of a continuing ban on all further dumping. It was the NUS boycott of sea dumping that brought about significant changes in the United Kingdom's waste management policy for low and intermediate level wastes with a switch from sea dumping to disposal on land. The Holliday Committee recognized the seamen's view of the sea as their 'back yard' and noted the depth of their feelings against using the seas as a rubbish tip for nuclear waste. While the British government and the nuclear industry all support an early resumption of sea disposal, mainly because future arisings of MAGNOX and AGR decommissioning wastes in the form of large concrete and steel components are too big for easy disposal in a deep mined facility without costly and hazardous cutting-up, it is now unlikely that the United Kingdom will be able to convince other nations of the safety and necessity of sea dumping.

## Dumping at Drigg

The Drigg disposal site is situated in West Cumbria about six kilometres along the coast to the southeast of BNFL Sellafield. It is owned and operated by BNFL and occupies about 120 hectares (300 acres). The Drigg site was once a Royal Ordnance Factory used for the manufacture and storage of high explosives, opening in 1939, but abandoned soon after the war. Authorization was granted in 1958 for the shallow burial disposal of solid low level radioactive wastes in trenches dug into the underlying boulder clays and disposal operations started in October 1959. Authorization was issued on the basis that the boulder clay would form an effective barrier against the contamination of the underlying St Bees sandstone aquifer by downward percolation of drainage water. So far only the northern half of the site has been used for disposal, while the remainder has been left wooded or is being used for storage, rail sidings and waste handling facilities. At present rates of disposal and if proposed improvements are carried out, it is expected that the Drigg site will remain in operation until the end of the century, when all the suitable areas of the site will have been used up. If the present rate of disposal is reduced by reserving the Drigg site for wastes produced solely at Sellafield and re-routing other wastes to another new site then the life span of Drigg may be extended another 30 years to 2030 (Grove and Hickford, 1984).

The annual volumes of waste disposed of at the Drigg site vary considerably from year to year. Over the last five years volumes have varied between approximately 30,000 and 100,000 cubic metres. In 1985 about two thirds came from BNFL Sellafield, 7 per cent from the United Kingdom's nuclear power stations, about 10 per cent from other nuclear establishments, 2 per cent from the National Disposal Service (NDS) (serving non-nuclear industries, universities, hospitals etc), and the remainder from other BNFL sites (BNFL, 1986). The wastes disposed of at Drigg typically include towels, paper, cardboard, plastic sheeting and containers, protective clothing, electrical cabling, scrap metal, process wastes and excavation spoil. Much of this waste, especially that from Sellafield, is potentially rather than actually contaminated in that it arises in the working areas of active plants and laboratories rather than from actual chemical processes (RWMAC, 1986). These wastes include 'surface-contaminated' wastes (non-radioactive waste materials whose surfaces are contaminated with a layer of radionuclides) rather than 'activated' wastes (non-radioactive materials such as steel which have been made radioactive inside nuclear reactor vessels).

The major contaminating radionuclides are strontium-90, caesium-137, caesium-134, ruthenium-106 and cerium-144. Important contaminating radionuclides in wastes from non-Sellafield sources are uranium, thorium and tritium. Plutonium contaminated materials (PCMs) above the authorized limits are not disposed of at Drigg, but some are stored on site in a specially constructed storage building. PCM is a term used to describe a large range of materials from disposable gloves to redundant machinery; all with varying degrees of contamination. Some PCMs were disposed of at sea, within the limits set by the London Dumping Convention, until 1983 when dumping operations ceased. At present PCMs are stored pending deep disposal. PCM wastes at the Sellafield site are treated (volume reduction), segregated into high- and low-plutonium content streams and drummed ready for storage at the Waste Treatment Complex (WTC) (Schneider, 1985). Some PCMs, which are not suitable for shallow land burial, are also stored on the site. The PCMs were originally stored in 200 litre drums and crates in disused explosive magazines. However, many of the drums corroded and some of the crates had broken up releasing their contents. This resulted in 'a great deal of contamination inside the stores' (Charlesworth, 1985, p.142). To recover and restore these wastes BNFL had to, at considerable expense, design, develop and commission a special facility capable of being sealed against the entrances of the magazines to allow access for workers (wearing protective suits) to carry out decontamination and repackaging operations without contaminating the environment. The facility had to provide full changing, monitoring and decontamination facilities for the workers, full conventional and contingency services, such as fire fighting capability and the ability to handle, monitor, repair and repackage damaged drums, and then withdraw from the area without leaving any contamination (Charlesworth, 1985). This recovery work was a painstaking and tedious task and has only recently been completed after 10 years. The PCMs have now been repackaged and are re-housed in a specially built store at the southern end of the site awaiting deep cavity disposal.

BNFL is authorized to dispose of solid low level radioactive waste at Drigg by the DOE and MAFF under sections 6(1) and 6(3) of the Radioactive Substances Act 1960. The site is also licensed by the NII under the Nuclear Installations Act 1965. The authorization came into effect on 1 April 1971 (DOE,1986; Hansard, 27 June 1986). A similar set of conditions was issued to the previous operators, the UKAEA, in July 1964. Although the materials disposed of at Drigg are always described as 'low level', Feates of the DOE in giving evidence to the Commons Environment Committee on Radioactive Waste, suggested that it is possible under the present conditions of disposal for individual packages to be outside the definition of low level waste. The conditions of disposal for Drigg are based on average levels of activity disposed of per day and not the activity of individual packages. Given the large volume of scarcely contaminated wastes, this overall daily limit could admit packages well above the definition for low level wastes.

In addition to the statutory conditions, BNFL imposes its own more restrictive conditions on the consignments of waste for disposal at Drigg. These require secure packaging and transport, adequate labelling, fire precautions and the exclusion of complexing agents, and include limits on transuranic radionuclides, radium-226 and tritium in the wastes, and limits on surface dose rates and contamination levels. Despite this, the Commons Environment Committee was still not happy about what they call the 'haphazard approach to what goes into Drigg' (Environment Committee, 1986), and suggested a number of improvements including: better monitoring of disposed wastes and the environment; better pre-disposal treatment of wastes; separation of contaminated and uncontaminated materials; better packaging and labelling; and categorization of wastes according to alpha and beta/gamma content and half-life.

The low level wastes have been buried in trenches about 6 – 8 metres deep, 25 metres wide and up to 700 metres long, dug into the glacial clays. Before a trench is excavated the ground is drained by digging a slit trench and connecting this up to the site drainage system. The trench is then excavated with its base sloping from its northern end at a rate of fall of about 1:500 to control drainage from the trench. Trenches are excavated by drag line to a depth of 6 – 8 metres depending on the local drainage features and the level and thickness of the clay layer (RWMAC, 1986). The sides of the trench are unsupported and stand at their natural angle of repose. During excavation of the trenches exposed geological features are recorded by a geologist. Where sandy areas are uncovered in the base of the trench, bentonite clay is spread on the trench floor and harrowed in to improve its impermeabilty. The waste is continuously tipped and buried with a minimum of one metre of soil which is then covered by a layer of hardcore (crushed stone), a layer of man made fibre netting to help provide stability and a surface layer of fine stone and ash. This is all compacted and used as a road surface over which waste skips can be transported to the tipping front. The typical radiation dose to a person standing on the finished trench surface is about twice the average background level from natural sources.

The geology and hydrogeology of the Drigg site has been investigated in detail using boreholes, investigation probes into existing trenches and surface water flow measurements. A number of borehole programmes have been carried out. Three boreholes were drilled along a centre line through the site as early as 1962, followed later by four holes in 1975. Between 1977 and 1981 the British Geological Survey (BGS) drilled 28 boreholes under contract. BNFL have themselves carried out two further drilling programmes for investigative and monitoring purposes and have since initiated a third. This is undoubtedly extremely sensible but it also implies that knowledge of the site was less than it should have been before dumping started and that major problems may well occur at some future date.

The geology at Drigg has been found to consist of a thick and variable sequence of sands and gravels, silts, clays and boulder clays overlying the St. Bees Triassic sandstone bedrock. The surface of the bedrock is very irregular varying from 10 metres below sealevel in the north of the site to 42 metres below in the south. The St. Bees sandstone is about 1000 metres thick (a borehole at Seascale penetrated 975 metres without reaching its base) and is an important aquifer in West Cumbria (Williams *et al*, 1985). The thickness of the glacial deposits at Drigg is variable, generally ranging between 15 to 45 metres and reaching a maximum of 58 metres. The deposits were formed by a complex series of ice advances and retreats during two glacial phases. This is reflected in the complicated nature of the Drigg deposits.

At Drigg the hydrogeology of the 'G5' boulder clay which is used for disposal is complex, due to the variable nature of the glacial deposits. This aspect is 'crucial' to operation of the waste trenches (Robins, 1980). Within the glacial sequence the clays (G1,G3,G5 horizons) represent relatively impermeable horizons which restrict the vertical movement of water, up or down. The intervening sand and gravel layers are more permeable and appear to represent localized aquifers which provide routes for both vertical and lateral flow of groundwater. The work of the BGS draws attention to areas of 'perched' water table within these deposits. The main water table is well below the floor of the disposal trenches, but in certain areas where permeable sand and gravels occur above the impermeable clays groundwater exists above the general level of saturation. Movement of groundwater in these 'perched' water tables could lead to 'small' amounts of radioactivity moving laterally away from the waste trenches (Jones, 1986).

BNFL have made investigations into the conditions within the trenches themselves and have discovered that the drainage at the northern end of the trenches is too slow. As a result the trenches have become flooded which provides a potential or 'head' to drive the movement of water in the perched water tables identified by the BGS. Evidence from geological information and flow measurements, for example, indicates that water flowing into the British Rail drain running along the northeastern boundary of the site originates from the flooded trenches via lateral flow through permeable sand layers (RWMAC, 1986). Since monitoring of the British Rail drain began in 1979, maximum levels of activity in the drain water have been approximately five times the derived limit for drinking water (RWMAC, 1986). One solution to this problem would be to pump out the flooded trenches and prevent further ingress of surface waters by covering the trenches with an impermeable cap. This is currently under consideration by BNFL as part of its proposed site improvements.

Lateral migration of radioactive leachate also occurs through 'valley-like' depressions in the top of the boulder clay surface and through drains in the waste trenches (Williams *et al.*, 1985). Rainwater which does not run off directly into surface drainage channels infiltrates downwards in the waste trenches. Drainage water from the trenches is routed via drains and culverts into the Drigg stream and off the site. This provides the most important pathway for the release of radiation from the waste trenches. Conditions of the authorization of the site were made so as to limit exposures which would result even if the stream were used for drinking water. Results from the monitoring of radiation levels in the stream suggest that the amount of radiation leached from the buried waste (as a fraction of the whole) is small. BNFL therefore maintain that the site is not an example of 'dilute and disperse' disposal as frequently suggested, but say that all but a tiny fraction of the radioactivity disposed of is retained in the trenches and that the small fraction leaving the site in the Drigg stream is acceptably low (Jones, 1986).

Radioactivity is monitored at the Drigg site both in the air and drainage water. Air samplers monitor the resuspension of activity into the atmosphere during the tipping of wastes at a number of points around the site. Activity in drainage water is monitored by the direct sampling of groundwater in boreholes, and the sampling of surface waters both in the British Rail drain and where the Drigg stream discharges from the site boundary. A preliminary assessment of the radiological impact of the site by the NRPB has indicated that the present and future impact of disposal operations is low (RWMAC, 1986). The NRPB therefore maintains there is little risk to the individual and that the highest 'additional' risk to the individual of developing cancer is about 1 in 10 million. On the basis of monitoring evidence and the conclusions of the NRPB, the RWMAC suggest that there is no need for any change in waste disposal operations, but nevertheless consider it 'desirable and sensible' for BNFL to continue with their studies and improvements (RWMAC, 1986, p.47). Currently, BNFL have in progress a programme of work costing £1 million relating to the future disposal operations at the Drigg site, in which 'urgent attention' is being given to to a number of improvements in disposal practices. BNFL also plans to spend another £20 million on improving the Drigg site for future use.

The Drigg disposal site has been used for the shallow burial of low level radioactive wastes since 1959. The site was chosen by the UKAEA, its original owners, purely on the basis of its availability and proximity to the Sellafield works. At that time knowledge of the hydrogeological conditions was somewhat sparse and so far no major problems have occurred, but it has existed for less than 40 years. Whether Drigg remains safe for the next 300 years or so depends on both the continued good performance of the older trenches which may be more a matter of luck than science, and also on the continued careful handling of future dumping. It is interesting that no opposition was voiced at the selection of the Drigg site, in stark contrast to the recent protests over NIREX's shallow land burial proposals. This is illustrative of the marked change in public attitudes and sensitivities towards nuclear matters that have occurred since 1959, with a shift from one of general apathy and disinterest to extreme sensitivity and lack of confidence. This has been brought about by an increasing awareness of nuclear issues through the rise in environmentalism and media coverage of a number of nuclear incidents such as the 1957 Windscale fire, Three Mile Island and Chernobyl. However, with Drigg it may also have been a reflection of a total absence of public information and participation in the original UKAEA selection and development process.

It is clear that Drigg is leaking! Drainage water from the trenches do contain leached radionuclides. This has resulted in Drigg sometimes being described as a 'dilute and disperse' site by its opponents. BNFL on the other hand stress that the fraction of the total activity disposed of at Drigg which is removed in trench leachate is very small, while the rest is retained within the trenches. However, removal of radionuclides from infilled trenches does occur by a combination of two processes; either by vertical percolation of water through the trenches and drainage into the Drigg stream or by lateral migration of trench water through permeable sand and gravel horizons in the form of perched water tables. Although the amounts of radioactivity released into surface drainage waters by these processes is apparently small, the fact that these processes occur at all so early on during the containment life-time of a radwaste dump would suggest that the hydrogeology of the site and the trench design is unsuitable by modern standards. Maybe BNFL's current programme will improve the situation in the near future. The point here is that clay disposal sites are clearly highly complex. Drigg appears today to be acceptably safe and is certainly possessed of extremely favourable economics, but the potential cost of having to exhume the site at some future date if really serious problems appear is likely to be massive. What better justification can there be in avoiding any need to rely on clay containment by seeking deep burial of all wastes regardless of nature and activity? As the PCM problem showed, what seemed to be a good idea in the late 1950s caused a major problem in the 1970s. It is extremely important with radwaste to make the right decisions at the first attempt.

**The 1983 Billingham site proposal**

The next sections provide details of those sites that NIREX have proposed as potentially suitable locations. It might be assumed that they are only of historical importance but it is fairly likely that at some time during the next 100 years they will again become the focus of attention either by NIREX (or its successor organization) or by other radwaste dumping agencies.

The Billingham site is probably one of the best in Europe. In both geological and economic terms the Billingham anhydrite mine was the natural and perfect choice as a candidate site for a deep repository for long-lived intermediate level waste which could be developed quickly and with minimal expense. The mine is owned by ICI. It opened in 1928 and closed in 1971 during which time 33 million tonnes of anhydrite had been removed. The anhydrite was extracted using the 'room and pillar' method which has left a honey-comb of 'rooms' separated by massive anhydrite 'pillars' at a depth of 140 – 280 metres and over an area of about 500 hectares. This method of extraction is about 50 per cent efficient and coupled with the strength of the rock (several times that of concrete) makes the mined cavity very stable and prevents any subsidence. The total capacity of the mined cavity is approximately 11 million cubic metres (ICI, 1983). The geology of the mine is well known from past mining activity. The main anhydrite seam in which the mine lies is about 6 metres thick and underlies great thicknesses of permain marl and shale sequences, which themselves are overlain by the Sherwood sandstones and surface deposits of sands, gravels and boulder clays (Bath *et al*, 1985). It is reported that the mine is very dry which is a major factor when considering sites for radioactive waste disposal and that the water flow into the entire mine is localized and only in the order of 13 litres per minute (Ginniff, 1985).

Locating a repository for intermediate level wastes in a stable geological formation deep underground such as the Billingham anhydrite is highly attractive since it reduces virtually to zero the chances of re-exposure due to natural or accidental disturbance (Ginniff and Phillipson, 1984). The major route by which radioactivity could be returned to man is by dissolution in groundwater which may eventually migrate to the surface. However, several barriers exist to limit water mobility and radionuclide migration in the near- and far-field environments surrounding the Billingham mine. The waste itself is very insoluble in alkaline conditions such as those which almost certainly occur within the mine given the alkaline nature of the anhydrite rock and the concrete in which the waste would be packaged (Saunders, 1987). Waste packages are designed to limit water access to the contained waste, forming another barrier to radionuclide release. Additionally, it is the nature of anhydrite that when it comes into contact with water it is transformed into gypsum (anhydrite being the anhydrous form of gypsum). The presence of anhydrite indicates that the strata have been dry for millions of years and are likely to remain so for millions of years to come or the anhydrite would have converted to gypsum long ago. There is therefore, very little water moving through the anhydrite strata in which radionuclide contaminants could be transported away from the repository. Should radionuclides find their way into the groundwater, both the anhydrite and the permain marls above are very strong adsorbers of dissolved radionuclides (that is radioactive contaminants are attracted by these rocks out of solution and become 'attached' to their surfaces) and thus make their effective rate of travel very slow. Potential sources of groundwater which may leak into the mine are from the Sherwood sandstone aquifer which outcrops above the anhydrite strata and the Brine cavities to the east which have been created by the extraction of salt using solution techniques (Morris, 1979). The combination of a slow rate of travel, the natural decay of radionuclides and dispersion into a larger body of groundwater make the return of radioactivity to man very unlikely.

Economically Billingham has also several advantages in its favour. The mine has good road and rail links with Sellafield, the principle source of intermediate level wastes and is also very near the Hartlepool nuclear power station (Ginniff, 1985). Suitable disposal cavities already exist in the Billingham mine and these would only require initial preparatory work such as road dressing and removal of surface irregularities before the mine could be used (Grove and Hickford, 1984). No major underground excavation work would therefore be necessary before the site was commissioned, although the possibility of alternative access from the Hartlepool nuclear power station site either to complement or replace the two existing shafts at ICI's Billingham complex might have to be investigated (Atom, 1983).

Existing tunnels could accommodate two layers of large concrete boxes with each being backfilled as and when appropriate. The size of the tunnels and existing access shafts effectively limit the size of larger waste packages to 3 metres (overall dimensions) meaning that large components from the decomissioning of nuclear power stations would require costly and hazardous cutting up before disposal at Billingham (Grove and Hickford, 1984). If the development of Billingham had gone ahead then NIREX envisaged the disposal of approximately 5,500 cubic metres of packaged long-lived intermediate level wastes per year up to the end of the century. To NIREX and the rest of the nuclear industry the Billingham anhydrite mine seemed the natural choice. The site is claimed to be unique in the British Isles (similar examples exist outside the United Kingdom, for example, the Asse salt mine in West Germany); it provides a ready made deep underground cavity in dry, alkaline, geologically stable conditions which is within convenient reach of the Sellafield reprocessing plant.

## The Elstow storage depot proposal

At the time of the announcement of their interest in the Elstow site, NIREX was looking for a clay site for the burial of low and short-lived intermediate level wastes. The Elstow Storage Depot lies in Marston Vale (the 'Brickfields') about

5km south of Bedford. The site is seemingly ideal being located in the clay lands of the Bedfordshire brick fields where the Oxford clays outcrop with considerable thickness. It is owned by the CEGB although it is at present partly occupied by 60 – 65 firms with short leases engaged in storage, industrial and retail businesses. The remainder of the site is given over to agriculture. The site totalling some 180 hectares was originally used during the Second World War for the manufacture and storage of ammunition, but was purchased from the MOD by the CEGB in 1969 with a view to developing the site for a new power station. However, the site is no longer required for this purpose.

The idea was that short-lived intermediate level wastes would be disposed of in deep, specially engineered and concrete lined trenches dug approximately 10 – 15 metres into the clay stratum. Wastes disposed of in this way would have to be isolated from the environment for about 300 years by which time their activity should have decayed to near background levels. This form of disposal is referred to as engineered trench disposal (ETD). Low level wastes similar to those dumped at the Drigg disposal site would be disposed of in simple unlined, soil-covered trenches approximately 6 metres deep, again similar to those at Drigg. This form of disposal is referred to as shallow land burial (SLB). Like the short-lived intermediate level wastes these low level wastes have half-lives in the order of 30 years and would also require isolation from the environment for at least 300 years, although heavy shielding is not necessary.

The Oxford clay at Elstow is of interest because it appears to have all those qualities which would have made it suitable for an SLB repository. The hydraulic conductivity of the clay is low and its absorption capacity is high. The surface drift deposits are confined to the western part of the site and would have been avoided by siting disposal trenches in the eastern half. SLB as a method of disposal is best suited to dry geologies where there is no groundwater to transport radionuclides away from the repository. Unfortunately no such conditions occur within the British clay geologies which are saturated for all, or at least part, of the year and it would be expected that the trenches would fill with water over a period of time, as have the surrounding clay pits. Containment depends, therefore, not on the absence of water in the clay but on the rate of water movement.

Regional groundwater flow in the Elstow area has been analysed, using available data, by the BGS. Elstow lies on relatively flat ground extending away from an abrupt scarp of lower greensand sitting on top of the Oxford clay. The greensand is a major aquifer and contains groundwater which because of its position above the Oxford clays exerts a potential for vertical flow into the underlying formations (Williams, 1985). Under these conditions it appears that groundwater flow from a repository would be downwards into the Kellaways sand favouring repository construction by creating a long time lag before groundwater movement would return radionuclides to the surface. The rate of water movement in the more permeable rock horizons below the Oxford clay and Kellaways beds can only be estimated so it is not known whether this time would be long enough for the radioactivity of transported contaminants to decay to safe levels. On the other hand the clays at the base of the Kellaways beds may constitute an impermeable barrier restricting downward movement of groundwater. It is likely that the limited thickness and conductivity of the Kellaways sands would limit horizontal flow at this barrier, resulting in upward movement of groundwater in the Elstow area. Such a situation would result in those radionuclides leached from a repository being returned to the ground surface around the repository in a relatively short space of time when levels of radioactivity might well exceed safety limits. Faults are known to exist within the Oxford clays and have been recorded in worked pits to the southwest of the Elstow site. These may form lines of increased permeability and water movement and may well complicate the pattern of groundwater flow. Borehole information for the Elstow site is sparse as to whether such faults exist; and, if they do, the effects they have are unknown. Future brick making activity in the area may also have an effect on the migration of groundwater from a shallow repository at the Elstow site.

Another consideration regarding the hydrogeology of the Elstow site concerns the chemistry of the groundwater. Both weathered and unweathered Oxford clays have a high gypsum content which it is thought has led to the high sulphate concentrations found in the water of flooded pits. This problem also occurs at Billingham and may have an influence on the integrity of the concrete used in repository construction and waste packaging. Sulphate-resistant concretes are available and these would have to be used throughout the repository. A more detailed description of the geology and hydrogeology of the Elstow site is given in Williams (1985) and Milodowski *et al* (1985). There are other possible problems in that the area around the depot has been extensively worked for clay to make bricks. Some of the clay pits are flooded and are now used for recreation: for example, the nearby Stewartby Lake. Other pits like the Elstow pit and 'L Field' are being used for the disposal of domestic and commercial wastes. The Oxford clays beneath the Elstow site are also considered by some to constitute a major brick-clay resource for the future, so use of the site for radioactive waste disposal would effectively sterilize these reserves and lead to a possible conflict of interests.

Finally, the site is good from an accessibility perspective. The A6 Luton to Bedford road runs to the east of the site, although the main access is via the B530 to the west which has a purpose-built slip road to accommodate the large volumes of traffic using the depot. The M1 passes the site some 12 km to the southwest. Rail access to the site is also good with the main Bedford to St Pancras line passing to the west. There are remains of the former depot rail sidings and it is probable that these could be re-built.

## The Bradwell proposal

The Bradwell site lies on CEGB owned land adjacent to the Bradwell Nuclear Power Station on the southern side of the Blackwater Estuary in Essex. The site is approximately 2km northeast of the village of Bradwell-on-Sea and 30km from Malden, the nearest large town. The site is also in close proximity to two designated conservation areas. These are the National Nature Reserves of the Blackwater Estuary and the Dengie Peninsula. Bradwell was one of the first commercial MAGNOX reactors in the United Kingdom, being commissioned in 1962. The CEGB-owned land adjacent to the reactor site has been reserved with the possibility of developing additional generating capacity. An SLB facility would not apparently preclude the building of a second nuclear reactor (NIREX, 1986b). This 300 hectare site was formerly part of an airfield and is presently in agricultural use. Access to the site is mainly via those roads serving the power station which link with the main trunk road network 35 kms away. The nearest railhead is about 10km away at Southminster. Access by sea would also be possible, having been used for the transport of materials during the construction of the power station.

The site lies on up to 15 metres of recent esturine and marine alluvium which overlie up to 50 metres of London clay. The London clay outcrops in the southwest part of the site, farthest from the sea, where the alluvial deposits are absent and have been replaced by a thin veneer of river terrace sands and gravels. The London clay overlies 25 metres of lower London tertiary deposits, consisting of sands and hard clays, which themselves overlie chalk. The groundwater in the chalk beneath the site is itself saline and no abstractions are made from this formation, but it becomes a valuable resource aquifer in the region to the west of the site (Robins, 1980). Robins (1980) maintains that the potential for the disposal of radioactive wastes in the London clay beneath the Bradwell site is 'poor'. This he maintains is due to its limited thickness and partial hydraulic continuity with the underlying chalk aquifer. More recent research suggests, however, that groundwater movements between the chalk and the London clays and tertiary deposits are 'almost certainly' upwards through the clays driven by the higher hydraulic potential in the underlying chalk (McEwen, 1986a). It is therefore deemed extremely unlikely that contamination of the underlying chalk aquifer could take place by groundwater flow from any repository built within the alluvial deposits and London clay at the Bradwell site. A more detailed description of the geology and hydrogeology of the Bradwell site is reported in McEwen (1985). The principal problem with the Bradwell location is its proximity to London. Additionally, it is also a potentially good location for further nuclear power plant both to replace the MAGNOX station and for additional power generation being well placed to serve the Southeast.

## The Fulbeck airfield proposal

The site at Fulbeck lies on a former airfield in Lincolnshire about 12km east of Newark-on-Trent. The site, which covers approximately 270 hectares, is owned by the MOD who retained the airfield after the war for use as a training area. Part of the site is leased as agricultural land. Access to the site is via a minor road which links to the A17 and A1. The nearest railway line is the main east coast line which runs 4km to the south. The site is relatively flat and lies between 13 and 16.5 metres above sea level on the gently west sloping lands at the foot of the Lincoln Ridge which rises abruptly to more than 100 metres in the east. The geology of the site is quite complex, being made up of a series of inter-bedded lower Lias clays, mudstones and limestones. The 'Obtusum-Oxynotum' clays outcrop at the site to a maximum depth of 20 metres. Beneath the lower Lias rocks is the Penarth Group that overlies the Mercia mudstone Group which is believed to be over 200m thick. Deeper still is the Sherwood Sandstone Group which is an important potable water aquifer to the west of the site although it is considered to be too saline for drinking water supplies in the Fulbeck area.

The regional groundwater regime in the Fulbeck area has been analysed by McEwen (1986b). Although there are several permeable limestone horizons within the lower Lias which may be called aquifers, they tend to be thin; as there is no evidence of any substantial volumes of water ever having been obtained from them, they can be assumed to be of little importance. Under the existing hydrogeological regime it is probable that groundwater movement is upwards through the lower Lias driven by the potential head in the higher ground of the Lincoln Ridge and the Sherwood Sandstone below. A more reliable understanding of the local groundwater regime is not possible given the sparsity of available data. Results from a detailed programme of borehole drilling would be needed before an accurate assessment of the site could be made.

## The South Killingolme site

The site at South Killingholme is owned by the CEGB and lies in a 'special industrial zone' on the south bank of the River Humber approximately 6km north west of Immingham. The site is in close proximity to the villages of East Halton and North Killingholme. The 330 hectare site is currently in agricultural use and forms part of a larger area of land for which the CEGB has consent for the development of a 4000 MW oil-fired power station. Plans for the power station have been shelved but 'alternative forms of (power) generation are under consideration and could be developed in conjunction with a repository' (NIREX, 1986e). It could presumably be a possible PWR site. Much of the land in the area is either agricultural or industrial. Where the land is well drained it provides good arable land and is used mainly for cereal crops. Intense industrialization has taken place on the low-lying areas bordering the banks of the Humber. There are several industrial estates, chemical works and oil refineries in the area and major docks at Grimsby and Immingham. The proximity to such a large concentration of hazardous industries gives rise to questions of safety similar to those voiced over the Billingham proposal although not perhaps as extreme.

The South Killingholme site itself lies on the flat, low-lying land between the

Humber and the Lincolnshire Wolds. Much of this land has been reclaimed from the Humber estuary and has a maximum elevation of only 4 metres above sea level. All drainage is towards the Humber. The geology of the site has been subjected to a number of site investigations carried out in the area for the CEGB, surrounding oil developments and the DOE. The site is covered to a depth of between 13 and 25 metres with a mantle of boulder clay (till), the intended disposal medium. This overlies chalk. Although the surface deposits are mainly boulder clay some Esturine and Marine alluvium overlie these close to the Humber estuary. There are two main layers within the boulder clay. The upper till, of stiff brown/grey mottled silty clay with fine gravel and chalk fragments has a maximum depth of 5.5 metres. This is underlain by a stiff brown silty clay again with fine gravel and chalk fragments but with the addition of fine sand horizons about 0.3 to 0.6 metres thick. The junction between the boulder clay and the chalk is marked by a layer of broken chalk pieces up to 10 metres thick. These are the result of ice action in the last glaciation and represent a layer of increased but highly variable permeability (McEwen, 1985b). The chalk below the South Killingholme site is the major water supply aquifer in South Humberside. The nearest abstraction of water from this is only 2 – 2.5km from the western boundary of the site. The importance of the chalk as a public supply aquifer means that 'containment of the wastes would have to be adequately engineered to prevent any contamination of the groundwater' (NIREX, 1986e). This factor alone reduces the inherent safety characteristics of the site.

The overlying boulder clays are relatively impermeable and tend to keep most of the water flowing along through the chalk from the recharge areas in the Lincolnshire Wolds to within the chalk itself. Discharge from the chalk does occur at a number of spring lines and there is some limited upward leakage through the boulder clays. The hydraulic potential within the chalk creates a hydraulic 'head' above the level of the chalk-boulder clay interface. This means that any water movement between the two geologies will be upwards into the boulder clay. In some areas, however, such as Grimsby where the rate of water abstraction now exceeds the rate of recharge from rainfall in the Lincolnshire Wolds, water levels have consequently fallen. The effect of this lowering of water levels is the ingress of saline water from areas of chalk under the Humber and possibly a reversal in the upward movement of water between the chalk and the boulder clay. For a repository to be developed at the South Killingholme site, careful control of groundwater abstractions within the area would be necessary to maintain the balance and prevent movement of groundwater through a repository situated in the boulder clay and into the chalk contaminating public water supplies. It would seem that this site is somewhat less than ideal unless the waste being dumped possessed a very low leachate potential; for instance, decommissioning wastes.

By contrast with the uncertainties of the clay geology, access to the site is excellent in all respects. A high quality road network serving the nearby Immingham Docks and associated petrochemical works runs by the site which provides direct links with the M180 and the national motorway system. Rail access is available via the freight line serving Immingham Docks which actually runs through part of the CEGB owned land. Sea transport is also feasible with deep-water mooring facilities at North Killingholme Haven and Immingham Docks. This site would seem to be well located for the disposal of slightly radioactive bulky decommissioning wastes and may well be revived in this capacity at a later date.

## A comparative assessment

*Table 5.2* Comparative strengths and weaknesses of the second round clay sites

| Advantages | Disadvantages |
| --- | --- |
| *The Elstow Storage Depot, Bedfordshire* | |
| – County planning policy had already accepted the possibility of a nuclear power station on site. | – Existence of substantial brick clay reserves beneath the site would lead to a conflict of interests. |
| – As a result of the above, no major housing or industrial developments had been planned within two miles of the site. | – Relocation of some 100 businesses would be required. |
| – Good road and rail access is available both sides of the site. | – The site is within close proximity to Bedford, a major centre of population. |
| *The Bradwell Nuclear Power Station, Essex* | |
| – The site is adjacent to the Bradwell Nuclear Power Station which has already become accepted by the community. | – The local road system is inadequate and the site is a long way from the nearest rail head. |
| – Sea transport of radwaste is a feasible option and an offshore jetty is available. | – The site is near two conservation areas. |
| – The site is remote and the surrounding area has a low population density. | – The area is used for recreation. |
| *The Fulbeck Airfield, Lincolnshire* | |
| – The possibility exists that military training could continue parallel to the development of a repository. | – The site is surrounded by and is itself high quality agricultural land and there is the possibility of an adverse effect. |
| – No dwellings on or next to the site and the density of the local population is low. | – Rail and road access are poor. |

The South Killingholme site, South Humberside

- The site is within an area designated for special industries 'not compatible with normal urban development'.
- Consent exists for the building of an oil-fired power station.

- Access is excellent and well developed in all respects.

- The proximity of certain hazardous industries raises the question of safety.
- A conflict of interests exists in that the land could be used for more special industries.
- The site is in close proximity to two settlements.

Source: NIREX (1986f)

A summary of the relative advantages and disadvantages of the four candidate sites is shown in Table 5.2. More important from a NIMBY perspective is the population distributions around the various sites. Table 5.3 provides a brief comparison. These results are also commented upon in Chapter 4. The fairly low close-in populations (say within 6–10km) around Drigg, Sellafield, Bradwell, Dounreay and Altnabreac contrast strongly with the very large populations 'near' to Billingham, Elstow and to a smaller extent to Harwell. The relatively low population counts near the nuclear sites of Drigg, Sellafield, Bradwell and Dounreay reflect the operation of the so-called remote siting strategy which was used prior to 1968 to locate Britain's nuclear power stations

*Table 5.3* Comparative population distributions around Britain's proposed and current radwaste disposal and storage sites

| Site Name | Distance Bands: 0–5km | 0–10km | 0–15km | 0–20km | 0–25km | 0-30km |
|---|---|---|---|---|---|---|
| NE Atlantic | 0 | 0 | 0 | 0 | 0 | 0 |
| *Drigg | 361 | 361 | 2,722 | 3,169 | 4,013 | 4,443 |
| *?Sellafield | 4,140 | 16,511 | 49,007 | 61,868 | 87,808 | 123,670 |
| *?Dounreay | 529 | 1,027 | 10,485 | 12,132 | 14,482 | 15,644 |
| *Harwell | 8,395 | 59,219 | 124,017 | 258,828 | 444,044 | 696,595 |
| ?Billingham | 113,963 | 320,960 | 438,828 | 587,561 | 743,230 | 891,283 |
| ?Elstow | 44,540 | 143,544 | 206,749 | 351,729 | 759,489 | 955,589 |
| ?Bradwell | 6,457 | 13,171 | 90,989 | 244,132 | 383,167 | 690,847 |
| ?Fulbeck | 2,788 | 25,156 | 81,878 | 189,798 | 281,006 | 380,755 |
| ?Killingholme | 7,119 | 35,007 | 295,402 | 547,682 | 602,984 | 681,275 |
| ?Altnabreac | 21 | 39 | 325 | 2,267 | 11,976 | 17,403 |

Notes:  * storage
          ? suggested radwaste dump sites

Source: estimates based on 1981 small area census statistics

(Openshaw, 1986). It is interesting that at a distance of 30km there is relatively little difference between the various sites, the exceptions being the 'really' remote sites of Dounreay and Altnabreac. The problem here is that there is not even an incredible accident scenario that could so affect a radwaste dump that populations at 10km, let alone 30km, would be at risk of anything nasty happening to them. Nevertheless, it is also easy to understand how people tend to consider that a radwaste dump is not a desirable neighbour or how it would do nothing to enhance the economic prospects of any region. It is also easy to imagine decades of 'cancer cluster' scare stories etc that can in no way be casually related to a radwaste dump but which may well cause massive and ongoing local concern. Proving the non-existence of a link is just as difficult as proving a link! NIMBY is a very real phenomenon and those silly enough to believe otherwise, should really think again. Just look at Billingham. Who could possibly have been so blinded by its apparent geologic suitability to overlook the 113,963 people within 5km (many living directly above it)? The northeast Atlantic seabed dump site avoids these people problems but creates others. Certainly there are no people anywhere near, but the radioactive stuff is merely dumped into the sea in the hope that it will be diluted and dispersed by natural processes. There is no other form of containment and no packaging to delay the release of the radionuclides into the sea water environment. The expectation is that the very large volumes of water will dilute it to acceptable levels and that foodchains will not concentrate it again. The experts may well be right, but it is all so very unsatisfactory.

## Comments on some of the counter-NIREX campaigns

It is useful to finish this site specific chapter by taking a closer look at the arguments that the County Councils Coalition (CCC), comprising Bedfordshire, Humberside and Lincolnshire County Councils, was marshalling to counter the NIREX proposals and to assess whether they would have stood any real chance of success had they been put to the test. It was noted in Chapter 4 that the County Councils had more research and staff resources available than did NIREX and that in theory, at least, they should have been able to develop an effective opposition to NIREX. Their success or failure would have been largely affected by: (1) the financial resources they could devote to the problem; (2) the extent to which they had explicit popular local support; and (3) whether they could identify key arguments indicating major flaws in the NIREX proposals.

Of the member councils of the CCC, Bedfordshire County Council had the longest time to prepare their case. Their opposition to the proposals was twofold. First, they were opposed in principle to the technique of the shallow land burial (SLB) disposal option. The County Council submitted evidence to the House of Commons Environment Committee in April 1985 attacking the shortcomings of the SLB concept by citing other countries who rejected SLB as a viable option and evidence from the United States where three sites in the wetter eastern states had to be closed because of problems with water infiltration, erosion of trench covers and leakage of radioactivity. However, these criticisms were not reflected in the subsequent BPEO report and no doubt NIREX could counter them by demonstrating a superior depository design.

The second line of attack concerned hydrogeological and environmental planning. The suitability of the Elstow site in geological and hydrogeological terms was viewed as questionable. The Oxford clay was chosen by NIREX as a suitable host geology for a near surface repository because of its impermeability, plasticity and high absorption characteristics. Yet the Oxford Clay, which stretches in a wide band from Humberside to the south coast, is at its thinnest at Elstow attaining a maximum depth of less than 20 metres. This is in apparent contradiction to the recommendations made in the 1982 White Paper, *Radioactive Waste*, which suggested a depth of 20 – 30 metres and the DOE's 1983 *Draft Principles for the Protection of the Human Environment* which took an upper limit of 30 metres for the minimum depth of the host formation. It has also been suggested that there are major gaps in the information pertaining to the hydrogeology of the site. Evidence from nearby areas suggests that the clay is laminated and not massive, and that it is subject to small scale faulting, while the water table lies above the unweathered clay in the four metres or so of callow and is underlain by permeable strata. In reality these geological deficiencies could be overcome by making changes to the depository design so they need not have been critical defects. In fact it could be argued that these problems would now be far better understood because of the 1986-87 drilling programme and the depository design could probably accommodate them.

A far stronger line of attack came from the local planning considerations. There were some suggestions that the site violated the demographic criteria NIREX claimed to be using. In the May 1985 issue of *PlainTalk* (the NIREX information newspaper) NIREX described the criteria used to define areas of suitably low population density as being 490 people per square kilometre or an upper limit of 100,000 people living within a five mile radius of a potential site. The County Council maintained that there were 120,000 people living within five miles of Elstow. Moreover, the site is near to south Bedford which is an area designated in the structure plan for future residential development. The weakness of this argument is simply that the population criteria have no statutory or radiological basis to them. They are purely guidelines. NIREX might have been expected to have argued that their Elstow facility would have been so safe that it would make no difference to public safety if the dump site was in Bedford itself.

The local development conflict was clearly a more promising approach. The DOE (1983) Draft Principles state that a site should be selected where it is unlikely that future development of natural resources or of the site itself will disturb the facility. Bedfordshire County Council maintain that it has long been their policy that the clay at the Elstow site should be used for brickmaking. It has been estimated that the clay reserves beneath the site could provide enough bricks to build half a million houses. Should development of a radiaoctive waste repository have gone ahead at this site then this resource would have been eliminated. To promote this use of the site as a viable alternative the County Council granted planning permission to the London Brick Company (LBC) to use the Elstow clay reserves to supply their nearby Stewartby brickworks. Although the LBC already had planning permission to exploit reserves near the village of Houghton Conquest, shifting extraction to Elstow would benefit both parties by providing LBC with a clay reserve four times the size of that available at Houghton Conquest, enough for 40-years' supply at Stewartby brickworks – and offer Bedfordshire County Council a seemingly more environmentally and publicly acceptable use for the Elstow site.

These arguments are also fairly weak. There is no national shortage of brickclay to justify reservation of the Elstow reserves in the national interest. Moreover, they do not own the site and presumably it cannot be developed by the LBC without the approval of the CEGB. These planning arguments would probably not have achieved very much at the Public Inquiry. To have any effect, a far more radical approach would have been needed; for instance, planning major new conflicting radwaste developments near to the proposed site or developing across all possible rail routes leading to Elstow. Neither seem to be feasible and unless some positive action is taken Elstow should be presumed to remain a possible radwaste dump site for use at some unspecified future date.

By contrast, Humberside County Council adopted a more intensive approach. In late 1985, the County Council altered their Structure Plan to give formal recognition to their opposition to the location of any disposal facility within Humberside. The reasons were: the lack of local involvement, credibility and impartiality of the government and nuclear industry, the technical uncertainty surrounding the concept of near surface disposal; and conflicting land-use and environmental considerations. In their Structure Plan Update they urged the government to reject NIREX's proposals to disperse the nation's radioactive waste to new sites and as an alternative to provide adequate above ground storage facilities at existing nuclear installations, allowing easy monitoring and avoid inflicting unnecessary risks on people in other parts of the country. To provide a strengthened basis for this amendment the County Council embarked on a 'wide-ranging but low-cost public participation campaign'. They distributed draft copies to over 500 groups and organizations for consultation and launched a publicity campaign. Response to the campaign was almost wholly in favour with all nine District Councils expressing their support for the Structure Plan Amendment. After considering the response in detail the final wording of the amended policy was: 'The County Council will oppose any proposals for the location of national radioactive waste disposal facilities within Humberside, including proposals for the exploration of any sites for such purposes.' As an anti-NIREX precaution, this is a useful step but one that was too late to have maximum impact. Had it been present earlier then NIREX might well have ruled out the Killingholme site. However, NIREX could argue that since it was *post hoc*, it has no legal basis, and constitutes only a minor hurdle. It would also have to be approved by the DOE who might well simply reject it. Yet the attempts that were made to mobilize public opinion and assess the level of opposition would have been useful but not massively so. Nevertheless, it does signal to NIREX the extent to which there is entrenched local opposition and that might have influenced any development decision, had the round two sites not been abandoned.

The final line of attack taken by Humberside County Council was also potentially useful. In their opinion the site at Killingholme was too important in terms of its potential for industrial use and the related creation of employment opportunities to be considered as a nuclear waste dump. They commissioned Coopers and Lybrand Associates, in association with Shankland Cox and Maritime Distribution Systems, in 1986 to assess the development potential of the Killingholme site. Their brief was to review and assess the site's perceived potential from a totally independent standpoint. They were not commissioned to assess the appropriateness of the site as a repository or to assess its potential in the light of the other sites under consideration. In their report 'Killingholme: the development potential' produced in October 1986, Coopers and Lybrand concluded that in their opinion Killingholme does have development potential for certain industrial land uses. They suggested that the site has potential as a port, with over a mile of river frontage and a large acreage of flat, accessible land. Although there is no shortage of industrial land in South Humberside, the Killingholme site is important because it represents the last major development opportunity on the Humber with good access to a deep water channel. At present maritime traffic on the Humber is increasing and so there could be some demand for new port facilities especially if designed to handle containers and RORO (roll on, roll off) traffic. It is estimated in the report that such a facility would require about 120 acres and would create around 350 jobs when completed.

Coopers and Lybrand believe that in the 'medium term' (up to the turn of the century) there is unlikely to be demand from a single major user such as a port. They do consider, however, that Killingholme represents a site of 'significant value in demand terms, from a specific group of development opportunities.' The remainder of the site not used by the port and its associated storage facilities could be used as the site for a new coal-fired power station. In assessing the site's benefits in the light of the economic and locational requirements of a power station, Coopers and Lybrand describe Killingholme as a 'premiere location', particularly in view of its accessibility to the Yorkshire and Nottingham coal fields, ease of access to the National Grid and port facilities should import of coal be necessary. The CEGB is currently planning to build three new coal-fired power stations and should Killingholme be chosen as the site for one of these it would create around 2000 jobs during construction and 600 jobs during operation. In association with a power station the site could house a substitute natural gas (SNG) plant used to produce natural gas from coal. By locating an SNG plant next-door to the coal-fired power station,

Coopers and Lybrand believe that considerable financial savings could be made from joint use of transport, storage and handling facilities. Although an SNG plant is unlikely to be commissioned before the end of the century, sites are presently being considered and it would be wise to reserve about 200 acres of the Killingholme site for an SNG plant. Such a development should it occur would provide about 600 additional jobs in the area. If the development of the port, the power station and the SNG plant were to go ahead, then it is likely that the remaining areas of the site could be made available to a number of new, related industries associated with the port facility and the utilization of by-products from the power station and SNG plant.

Coopers and Lybrand believe that these are the most 'realistic' development options for the site, but all would rest on the agreement of the CEGB who own the land. The CEGB stated at the time, however, that a nuclear repository would not be an unacceptable neighbour to a coal-fired power station and both could be developed at the Killingholme site. Should the site not be selected for one of the three new coal-fired power stations planned for the late 1980s and early 1990s then the site would probably not be considered by the CEGB for developing extra generating capacity until 2005. In this case Coopers and Lybrand suggest use of the land until that time for the port facility and a 'trade park' of associated industries while small areas could be used by 'difficult neighbour type' industries in the petro-chemical sector. It is admitted that this would require a substantial input of public resources and 'positive and proactive site promotional/development effort'.

As part of their research, Coopers and Lybrand interviewed a number of businesses and organizations involved in the South Humberside economy. Although a number of respondents did not believe that any development by NIREX would adversely affect the local economy, a 'considerable number expressed the view that NIREX could severely restrict the opportunities for further development, not just at Killingholme, but throughout Humberside' with a blighting effect. Building societies and estate agents in the Humberside area also expressed their concern that property prices in the Killingholme area might suffer as a result of the NIREX proposals. Evidence had already shown that the short listing of the site had caused some unease in the property market. This unease was quoted by the Northern Rock Building Society as the reason for Barretts pulling out of a house transfer deal, while the turnover of property in the Killingholme area had stagnated.

Some of these arguments are of considerable value. The planning blight approach is not particularly worthwhile. By contrast, if the radwaste dump could be proven to be denying strategic port development opportunities to the entire Humberside economy then there would be a case against the radwaste dump. On the other hand, it is doubtful whether the Coopers and Lybrand report was sufficiently firm to prove this case. The development potential was seen to be 20 or more years away and was dependent on many 'if's. Additionally, NIREX might well have argued that their radwaste dump proposal would have in fact acted as a catalyst for these developments and brought them forward in time. There might even be a case for supposing that these positive aspects, and the spin-off employment they would have created, would outweight the dangers of nuclear blight. So again, it would seem that the County Council had failed to develop a totally convincing defence. With site ownership in CEGB hands, virtually the only anti-dump policy that would virtually guarantee success would be to encourage incompatible industrial landuses to move into the area around the site, thereby increasing its unsuitability and restricting any further development potential. The time to take this action is before the national need for a seaport based, bulky-object decommissioning waste burial site is re-discovered.

The response of Essex County Council to NIREX's proposals for the Bradwell site were muted by comparison. Essex did not join Bedfordshire, Humberside and Lincolnshire in the CCC because its grievances about the proposals were apparently different. Essex County Council maintained that the development of a near surface disposal facility ran counter to their continuous policy of a progressive reduction of the import of hazardous waste into the county and that such a development would seriously affect their commitment to the promotion of employment and tourism within Essex.

# 6

# How NIREX select sites for a radwaste dump

A key question is how do NIREX identify potentially suitable sites for radwaste dumps? For example, why on earth was Billingham or Elstow picked? It certainly was not a random choice, even if some of the residents may have thought so. If you want to criticize what NIREX do then it is important to understand what it is that they do. The traditional nuclear engineer's view of the siting problem is that the primary criteria are purely technical, factors that are related to the safety, design and cost of the proposed development. Anything else would seem to be incidental. Sites are only invalid if they fail these technical criteria. Matters of public acceptability are basically irrelevant because the objective is to build a carefully engineered disposal facility which in order to be built must by definition be completely safe. It was this style of thinking that resulted in the selection of Billingham. It was this sort of technocratic approach that has also characterized the siting of nuclear power plant (Openshaw, 1986). It is quite logical and can be justified but it also encourages myths and blatantly incorrect statements about the location of nuclear facilities. For example, in *PlainTalk* (1984) NIREX responds to the question, 'Why don't you store or dispose of waste away from areas of dense population?' by declaring, 'It is very difficult in the United Kingdom to find areas that are remote from population and also have the geological attributes required for a radioactive waste repository' (p.8). This can only be verified by performing a detailed analysis of geologically suitable areas overlaid with population characteristics. Later in this chapter, and also in Chapter 7, it is suggested that there are potentially quite a number of seemingly suitable locations away from areas of dense population. So it is argued that NIREX are promulgating an old CEGB myth that you cannot have remote siting in the United Kingdom because there are too few suitable remote sites. This view is equally incorrect in both contexts and is even more misleading when applied to radwaste dumps because their selection criteria are probably less stringent than would be the case with nuclear power stations. There are grave dangers in adopting such a blinkered engineer's view of the world and thereby implicitly using a siting stategy which is characterized by a type of narrowly focused technical determinism to be operationalized using old-fashioned manual rather than computer based methods. The world today is a far more complicated place in which 'soft', intangible but very real psychological and socio-economic factors will intervene through non-scientific channels and political systems to frustrate completely any naïve technical view of reality. After Chernobyl not many people are going either to believe or trust anything the nuclear industry says, even if it is in fact true!

When the first sites were announced in 1983 there was very little detail given about a siting process which seemed to be a classic implementation of the traditional nuclear engineer's view of the world. No details were ever published of the list of sites that NIREX claimed to have evaluated prior to the final selection of Elstow and Billingham. The issue seemed to be regarded as a purely commercial matter and therefore immune to public requests for details. *PlainTalk* (1984) declares, 'We [NIREX] are a commercial organisation and deal with other commercial organisations in normal business confidence' (p.8). However, it was not too surprising that NIREX soon felt the need to describe in general terms how they arrived at their selected sites. As a result NIREX now have a number of publications detailing their site selection procedures. It may appear that these are partly an attempt at *post hoc* justification and that they were not in full use during the selection of what we term the 'Round-One' sites (Billingham and Elstow). Indeed they seem to have been developed in the search for alternatives to compare with Elstow, starting in 1985. Anyway, there is a now a requirement to have an explicit siting procedure (DOE, 1985) and NIREX obviously have to conform to this guideline.

## The DOE's siting guidelines

The task of NIREX to find a site(s) for the disposal of low and intermediate level radioactive wastes has been made difficult from the onset by the lack of a comprehensive and coherent set of guidelines to fully operationalize the government's waste management policy. The draft disposal guidelines published by the DOE in 1985 stated that in the evaluation of alternative sites the developer (NIREX) '...will be expected to show that he has followed a rational procedure for site identification'. However, the guidelines continue '... he will not be expected to show that his proposals represent the best choice from all conceivably possible sites' (DOE, 1985; p.19). Earlier versions of the same document appeared more rigid. The advanced draft submitted as evidence to the Sizewell Inquiry in 1984 had stated that the developer must demonstrate that, in selecting a preferred site, better site options for limiting radiological risks had not been ignored (DOE, 1984). The original consultative version also emphasized town and country planning criteria and policies in this context. These changes may be seen as a deliberate de-emphasis of the importance of site comparisons and the evaluation of alternatives (Openshaw and Fernie, 1986). Why these particular changes should have been made by the principal and independent government department with responsibilities in this area is something of a mystery.

Nevertheless, this single change of emphasis has allowed preferred sites to be forwarded without a comprehensive and rational evaluation of the alternatives and with no proof that even a relatively 'good' site has been found. There is a strong case for suggesting that as a matter of good siting practice any major developments which rely on national interest arguments really require a level of comparative site evaluation which is far more rigorous than the current DOE recommendations imply. When massively unpopular but also tremendously significant developments are being proposed the onus really must be on the developer and not the protester to demonstrate in an explicitly rigorous and scientific manner that the 'best' site and not merely a 'feasible' site has been found (Openshaw, 1988). At the Sizewell PWR Inquiry this task was considered to be a matter for the objectors rather than the developer (Layfield, 1987). In the case of nuclear waste, justification of the chosen site will require a more rigorous siting evaluation than in previous nuclear inquiries. The reasons for this include: (1) there is no benefit from the development; (2) there are currently only plans for one facility; (3) the national interest argument is overwhelming; (4) whatever site is selected there will be public concern and opposition; (5) there can be no good reason for NIREX not wanting to find the 'best' site and, having found it, prove its superiority; and (6) the technology now exists that makes it possible to perform automated computer searches of all candidate sites in a broadly based screening process that is locationally comprehensive.

The DOE have established the basic principles that will be used to assess in general terms any new land disposal proposals. They state (DOE, 1985, pp.16–17) that: (1) the site should be chosen and the facility designed so that the risk of cancer to any member of the public is not greater than one in a million in any one year; (2) any future movement of radioactivity from a facility should not result in a significant increase in the natural background activity; (3) a site must be selected so that it is unlikely that any future development of natural resources or of the site will disturb the facility; (4) there must be adequate provision for environmental monitoring including a benchmark against which to measure future performance; (5) the proposed treatment and packaging of waste and its assignment to a particular facility must conform to the national strategy developed by DOE; (6) there must be adequate provision for detailed record keeping; (7) there must be evidence that, using established technology, the facility can be closed and surface installations removed – as well as adequate funding for these operations; (8) the necessary arrangements must be made for post-closure institutional control, including the preservation of details of the facility and records of the type and location of waste.

In considering the radiological protection objectives, the DOE recognize a distinction between the institutional management period, which can be divided into pre-closure period (likely to last several decades) and the post-closure period of institutional management (which might last a few hundred years); and the post-institutional management phase during which it must be established that on-site supervision is no longer required. All very sensible. It is pity, therefore, that the DOE's siting guidelines were published and finalized after the first round sites had been selected! These guidelines also suggested that three alternative sites should be examined for each type of facility. However, the guidelines were little more than statements of general principle, containing no details of specific siting criteria. The relevant factors of geology, hydrogeology, population distribution and accessibility are merely listed, but their interpretation is left to NIREX. Table 6.1 lists these siting factors. How these siting factors are to be used is again left open with no guidelines as to how numeric evaluation criteria may be defined and then applied in real siting decisions. However, there was one major effect in that once Billingham was rejected, seemingly because of a major petition to the prime minister and the apparent political indefensibility of an urban site for a nuclear waste dump, a second attempt had to be made. It is, indeed, a measure of the massive gap in the levels of social and political understanding of nuclear scientists that so few seemingly considered an urban site unacceptable for a waste repository. There was after all nothing for the residents to worry about. This view may well be technically correct, but it is also absurdly irrelevant and incredibly stupid to put such faith in purely engineering arguments that are impossibly complex and wholly incomprehensible to ordinary people; and then to expect that this would be sufficient to overcome the resulting phobias, fears, nuclear neuroses and

*Table 6.1* Factors to be considered in the environmental assessment of radwaste sites

| Heading | Details |
| --- | --- |
| Technical | physical characteristics of the site |
| | nature and quantity of waste to be disposed |
| | hydrology |
| | geology |
| | climatic |
| | air pollution |
| Countryside conservation | agriculture and fishery uses |
| | protected areas |
| Socio-economic | employment impacts |
| | housing impacts |
| | tourism and recreation impacts |
| | emergency facilities |
| Planning | demographic |
| | transport |
| | proximity to other uses |
| | implications for structure and local plans |

Source: DOE (1985)

ignorance of a whole town. This public relations disaster demonstrates that political support for a nuclear repository is seemingly restricted to the general principle that there should be one somewhere, rather than for a particular site.

Not surprisingly the DOE might be regarded as being somewhat kindly disposed towards NIREX. They rejected the Environment Committee's attempt to develop a more rigid set of siting criteria and that the department itself (rather than NIREX) should be responsible for the final short-list of sites because of the large national interest stake that is involved (Environment Committee, 1986; Recommendation 36). The government's (that is DOE's) response was that, 'To go further [than the DOE 1985 document], as has been suggested, and specify wider criteria covering, for example, the relevance of communications or proximity to centres of population, would be inappropriate. NIREX, as the developer, must be given sufficient scope to deploy their own expertise within the legal constraints which provide comprehensive mechanisms for the scrutiny of any proposal for development' (DOE, 1986a, p.26). It should be remembered that if NIREX fail or default, then the DOE would have to pick up the challenge and they obviously do not wish to become that involved in such a major, no win problem. However, surely that is what government department's are for!

## The IAEA guidelines

The basic site search process adopted by NIREX is based on an International Atomic Energy Agency publication (IAEA, 1983). This source does not of course imply any external validation of the process only an indication of how best to proceed. It might, however, be difficult or embarassing if there were to be a major departure. The IAEA guidelines suggest that when searching for a rock cavity for LLW and ILW then the following three stage process is appropriate:

Stage 1: National survey and evaluation,
Stage 2: Site identification, and
Stage 3: Site confirmation.

It is difficult to imagine how else a rational site search process might be configured. Stage 1 involves defining areas that offer potentially suitable geologies and which are also feasible when judged in terms of transport logistics, planning issues and relevant socio-economic factors. This might be regarded as the total set of potentially suitable locations. In Stage 2 the feasible sites are narrowed down by considering aspects such as site availability and by more detailed desk studies of their suitability. This results in a small number of potentially suitable sites being identified. In Stage 3 the small number of sites identified by this preliminary process are thoroughly investigated by geological and geophysical means to confirm their suitability. This would be followed by a decision to develop one or more of the selected sites.

NIREX claim to follow this procedure (NIREX, 1987) no doubt moderated by the need also to satisfy the 'green book' (DOE, 1985). The DOE stress the importance of adequate containment either by the nature of the geological host rocks or by engineered safeguards, or by both (DOE, 1985, p.11). They stress the importance of three sets of site-specific factors that must be considered fully in justifying the choice of a site. They are: topography, geology, hydrogeology and hydrology (with special regard to radionuclide movement or retardation, and the long term stability of host rocks); demography; and the implications for the transport of wastes from places where they are generated or stored. All this is eminently sensible, rational, and hardly surprising. However, the process of identifying and evaluating sites is not defined. For instance, in the IAEA procedure, it is not clear as to what the suggestion 'survey the country' means. Does this mean detailed, high resolution studies of geology, transport routes, population density, etc using all the best available data resources? Or would a quick look at a national map also qualify? There is a feeling that maybe the original round-one sites were selected too quickly after NIREX was created to have allowed sufficient time for an in-depth survey given the nature of the available manual map analysis and overlay methods that are likely to have been used.

Another possible criticism concerns the temporal sequence of the stages. The CEGB in siting a nuclear power station sometimes appear to work backwards; here is a highly favoured site and the task is to develop a justification for it, perhaps by considering that the logically prior but missing Stages 1 and 2 have been implicit in previous site searches and investigations and therefore do not need to be repeated. Indeed there is a potential saving of expense and time by working backwards. It is quite legal because there are no rules as to how sites should be selected and evaluated in Britain. Often it appears that attention is mainly focused on the suitability of a previously chosen site taken in isolation, or with very limited comparisons being made. Once round-two attempts began, with DOE guidelines specifying the need for an explicit site evaluation procedure, NIREX seem to have developed their current site search procedures. The delay before the announcement of these new sites is probably a reflection of the complexity of the task. It is, however, still not clear as to whether it was a genuine 1–2–3 stage process and not a 3–2–1 approach that was used. It must have been very tempting to start by testing out a list of sites held by nuclear industry friendly organizations, while also performing a IAEA-type of search. Other interesting points involve the decision making used to refine the purely deterministic environmental definition of geological feasibility in Stage 1 and the gradual incorporation of much softer and fuzzier planning and socio-economic factors. Certainly, by Stage 2 the opportunities for *ad hoc* decision making and evaluations increases enormously. Ideally, this process should be documented and submitted as part of the evidence that a rational siting procedure was in fact applied.

Figure 6.1  Areas of suitable geology for round two sites

## A NIREX siting example

The most detailed published account is that given by the deputy director of NIREX in their newspaper *PlainTalk* (1985). An article entitled 'The meticulous five-steps in the search for a site' (Beale, 1985) describes the process that NIREX use. This offers an apparently comprehensive everyman's guide to repository site selection from the first basic steps to the identification of alternatives for detailed study, leading to the final choice via a public inquiry. The basic phases in this operation are: (1) geology, (2) population, (3) conservation, (4) accessibility and (5) areas of search.

The first stage involves mapping on a national scale those areas of Britain which seem to be most suitable for a disposal site. For the round-two sites, published geological maps were used to identify the principal clay outcrops. A digital version is shown in Figure 6.1. Geology often appears of prime importance in radwaste site searches because it provides an extra layer of containment. As Beale (1985, p.4) puts it, 'Using this belts and braces approach, geology becomes an important factor'. Should the man-made packaging decay and the engineering structures fail, as they will with time, then the geology provides an additional containment. Clays are a preferred geology because they have low permeability and they have some capacity for filtering out and retaining many of the radionuclides involved. Contrary to expectation clay environments are not usually water free but the water moves through them very slowly. At Drigg it is the clay itself, rather than the waste packaging or any other form of engineered containment, that has to contain the radioactive waste for sufficiently long periods of time. Drigg might therefore be regarded as somewhat crude. For the round two sites more elaborate concrete linings were being proposed in addition to the clay containment; indeed, it is now possible to provide containment purely by engineering means. The clay is now regarded as the second part of a multi-barrier method of disposal whereby a number of barriers are placed between the radwaste and the environment occupied by man. It is not clear as to why clay containment matters if, as is claimed, the facility can be engineered to provide containment by itself, although it is clearly a sensible step and it explains the overriding interest in geology. Another purpose may also be that of economy; the presence of clay containment may well allow a cheaper and simpler depository design to be used. A modern radwaste dump should not have to rely purely on geological containment, especially if it is near the surface.

The next factor is population density. This is clearly important because it directly affects the strength of the 'not in my back yard' (NIMBY) phenomenon. NIREX claim that proximity to large numbers of people is irrelevant because the waste repository design will be so safe as to convey no significant additional health risks. Yet Beale (1985) comments, 'In practice, however, given a choice, areas with low population densities are to be preferred to minimise the risk, however small, of any adverse impact' (p.4). There is a contradiction here. If the repository is so safe how can there still be any

*Figure 6.2* Areas with a population density greater than 490 persons per square
        kilometre

sufficiently large risk to justify a preference for low population sites? If the argument is that this is purely an additional degree of safety, then the obvious response might be, 'why is it that unsafe?'

Another problem is identifying a suitable set of population density limits. The nuclear installations inspectorate (NII) have a set of maximum population limits which they apply to current AGR (and PWR) reactors (Charlesworth and Gronow, 1967). These relaxed demographic siting criteria have allowed both the Hartlepool and Heysham nuclear power stations to be built in semi-urban areas. It now seems that these same limits are also appropriate for a radwaste dump even if there were no logical scientific basis for them in the first place. They represent no credible radiation accident, so why are they still being used? Anyway, this is not really a NIREX problem. Beale (1985) merely converts the maximum population of not more than 100,000 within five miles of these semi-urban reactor sites into a population density threshold, he writes, 'This represents a population density of two people an acre, and is the figure taken as a threshold for the purpose of taking a national overview in identifying possible repository sites' (p.4). It is a pity that there are approximately 164,000 people living within five miles (8kms) of the Hartlepool AGR site, indeed the nuclear installation inspectorate's relaxed population criteria would permit up to 176,000 people within five miles (7kms) (Openshaw, 1986). This would suggest that Beale's population density threshold of 490 persons per square kilometre needs to be almost doubled to 870 to be consistent with its claimed origins. Alternatively, if the NII's 20 mile site population limits are used, then this would imply a maximum density of 7.7 persons per hectare. Figures 6.2 and 6.3 show those parts of Britain excluded on grounds of too high a population density according to two variants of this maximum population density approach. Clearly NIREX have a possible problem here. The best they can hope to do is to utilize the current NII relaxed siting demographic criteria as a guideline. If this is the purpose, then their current population threshold needs to be almost doubled in size and is therefore of considerably less restriction that it would otherwise be. On the other hand, the authors do believe that it is really very sensible to use more strictly defined criteria and would prefer even more stringent thresholds to be applied over a much larger distance.

If the objective in having demographic limits is really to reduce potential NIMBY effects then a siting strategy incorporating remoteness from people really needs to be considered. In Britain there has always been a problem in defining remoteness given the nuclear industry's belief that remote sites are extremely rare. As a result there has been a tendency to consider remoteness in terms of either five or ten mile distances. Quite frankly, this view is convenient rather than accurate. Remoter sites, in reasonable locations, certainly do exist and would probably have been used for nuclear power stations had there not been conflicts with other seemingly more important siting criteria; for instance length of new transmission line, land ownership and a poor attempt to avoid areas of high amenity and landscape value. These same beliefs appear to have been transfered to NIREX. Yet there is an opportunity here to assess the

*Figure 6.3* Areas with a population density greater than 490 persons per square kilometre

extent to which greater levels of public acceptability can be 'bought' by deliberately seeking remote sites which are well within the constraints suggested by estimated health risks. The costs would be shared both by NIREX and the nation. The landscape, planning and amenity implications for example would not necessarily be negligible, but it is argued that they would probably be small (compared with oil related developments) and it would in any case be a trivial price to pay for public acceptance. It should not, therefore, be excluded on a priori grounds.

Radwaste dump siting cannot simply be regarded as a matter of picking a site and then ensuring that national safety limits are satisfied by repository design. This sort of thinking leads back to Billingham and the utter impossibility of persuading people that living above, or 'near', a radwaste dump is (a) harmless and (b) good for the neighbourhood. It would be better to grasp the stark realities of the problem and operationalize a strategy that seeks to minimize the population within say, 30 – 50 miles of a site, 30 miles being a pragmatic approximation to an 'out of site, out of mind' criteria. A 50-mile population minimizing approach would be better still. It is important, as has been explained elsewhere, because the sites have to be acceptable now and then remain acceptable for the indefinite future when public attitudes may well be different. It is easy to assume that better education will lessen the problems, but there is also the growing risk of greatly increased levels of public sensitivity to all aspects of the nuclear industry. This may be seen as resulting from (1) better knowledge of what is going on; and (2) nuclear accidents, some probably catastrophic, elsewhere in the world. There is no point in building a radwaste repository only for it to be closed, and probably exhumed at massive expense, ten years or so after it starts operations. Since NIREX should be seeking to use all reasonably practicable steps to ensure public safety and long-term acceptance, remote siting more than anything else (assuming a safe design), offers considerable benefits in this area.

Returning to NIREX's present siting approach, the next step is to remove areas which have been designated as being of national importance in terms of nature conservation and landscape protection. Figure 6.4 shows where most of these areas are located; further details are given in chapter 7. This reflects NIREX's lack of compulsory purchase powers as well as the usual attempt to have due regard for amenity and environment. It involves excluding from the national search such areas as national parks, areas of outstanding natural beauty and heritage coastlines. It should be noted that this really is a new departure for the nuclear industry in that on some previous occasions these landscape designations have been ignored in the national interest. It would be extremely mischievous to suggest that sites of low population accessibility only occur in such areas. However, it is also clear that developments in areas of natural beauty and recreational value would be opposed. The Countryside Commission sets out its policy as follows: 'It would be inconsistent with the aims of designation to permit the siting of major developments in protected landscape areas and only over-riding national interest and the demonstration beyond all

*Figure 6.4* Areas where landscape amenity and other exclusion factors may apply

reasonable doubt that alternative locations for the development are not available should justify any exception.' (ERAU, 1988, p.35). No doubt a sufficiently powerful justification could be made, if need be. National parks already contain some large-scale developments and there is a strong case for choosing these if the geology is right and the sites are sufficiently people remote.

A final factor is that of accessibility. Beale (1985) writes: 'If a choice of equally qualified sites were available, the preferred location would seek to minimise transport distances from the sources of waste to the site. This has a significant effect on transport costs and reduces the chances of transport accidents' (p.5). This appears reasonable but it is in fact a highly complicated procedure that is better applied when the set of all feasible sites have been reduced by applying other more deterministic criteria first. It would also seem daft in 1989 to pick a site purely because of its high accessibility to radwastes based on present day transportation costs. Modelling transport costs requires knowledge of present and future radwaste arisings for the duration of the repository's life. It also requires knowledge of transport costs over an extended period of time. Finally, if safety is a factor that is more important than transportation costs, then some regard needs to be given to the population living along the radwaste transportation corridors. There is a potentially large fraction of Britain's population who, without realising it, live near to one or more radwaste transportation corridors. This constitutes a large hidden reserve of possible future anti-nuclear protesters and the potential risks they present need to be reduced by the most careful manipulation of the siting process to minimize the numbers who may feel themselves threatened at some future date. Minimizing NIREX's operational costs is not really a valid objective given the presence of far more fundamental problems that threaten repository viability and not just its profitability. In a monopoly situation, transportation costs should not be such a significant factor. Here Figure 6.5 shows details of a rail and principal road buffer 3km wide. These areas might be seen as being locations of easy to develop sites from a transportation point of view.

The four factors (geology, population density, conservation and accessibility) are then combined to identify areas which show the greatest potential for near surface repository sites. It is a classic sieve mapping approach. It does provide an information base by which to screen sites and to identify prospects for further investigation. This area-based approach also has to take into account additional factors such as planning issues, mineral resources, and quality of agricultural land. In addition, it seems that other possible sites that were known to exist could be added to the list if they met the basic requirements. Beale (1985) comments, 'This does not mean that a potential site cannot be located outside these areas' (p.5). The basis for this statement is not given and it is puzzling. If the search process was comprehensive then there should be no additional sites that can be added in an *ad hoc* fashion, unless the criteria have been changed. It is difficult to reconcile *ad hoc*ery with a supposedly rational approach unless the results of the sieve mapping exercise are of such poor quality and are far too overgeneralized to be of much use as a means of

*Figure 6.5* Areas within 3km of a railway or major road

identifying particular sites. It is certainly much easier to start with an existing list of possible sites (namely, land owned by the NIREX partners but not yet developed) and then screen them for acceptability. The question is whether or not this would amount to a rational siting process in terms of DOE (1985). The answer is probably 'yes'!

The final step is a kind of 'narrowing down' process. Once the potential sites, or more correctly a small subset of all such feasible sites, have been identified then the real selection process begins. This involves more detailed reviews of geology, hydrogeology and availability, although it might be imagined that these criteria have already been used once to identify the candidate sites before this more detailed evaluation stage started. Anyway a list of sites are drawn up on this basis and further more-detailed, site specific, investigations are performed. They include: visiting the sites, updating the available information, and making a final assessment of suitability on the basis of three major factors: (1) safety (geology and hydrogeology); (2) planning (population density, accessibility and land use); and (3) technical feasibility (site access and constructability).

Specialists are then asked to score the performance of each site in respect to these three factors. Since some factors are more important than others, the factors were weighted. Safety is considered to be the most important, followed by planning and technical factors, and the sites are ranked accordingly in order of suitability. According to NIREX the four sites eventually chosen for detailed field investigations were those which gave the greatest priority to safety and met all the other requirements. How this scoring and weighting approach was performed, and what minimum requirement constraints were set, is a mystery to us. It could only be arbitrary, with different experts obtaining different results. Such uncertainty could be handled only by using fairly sophisticated multi-criteria evaluation methods. It is not known what NIREX did or used.

A major weakness here is that the optimality of the final selections is dependent on the set of sites from which they were originally selected. This candidate site list might not necessarily provide a representative sample coverage of the universe of all potentially possible sites. There is no way, therefore, of being able to determine 'how good' the preferred sites really are or how many 'better' or 'equally good' locations exist in Britain. It can be argued that these criticisms are not relevant because they go beyond what the DOE siting guidelines require from NIREX. However, these DOE guidelines should perhaps be regarded only as indicative of what is required and are not really a minimum standard. It is clear that the NIREX task would be complicated by their need to be able to demonstrate that very good or nearly optimal locations have in fact been found, thereby justifying their use as radwaste dumps in the national interest. It may be imagined that proving 'near optimality' rather than the far simpler task of proving feasibility, might well be a feature of any future Planning Inquiry. We would strongly argue that the preliminary site selection stage is recognized as being of absolutely crucial importance. The strength of the final case is critically dependent on knowing how well the preferred sites rank in terms of a universe of all potentially feasible sites.

## The 'Way Forward' era of siting

The abandonment of the four round two clay geology sites in 1987 would seem to vindicate the criticisms that have been made of the siting procedures. Nevertheless it seems that the siting procedures have not really changed, even if the target geologic criteria have. The greatest change is a switch from a near surface shallow land dump for LLW to a deep cavity disposal site for both LLW and ILW. It is obviously an improvement to have a single facility that is maximally safe by being placed far below the surface. On the other hand, these advantages are offset by the storage of ILW, which require to be contained for geologic rather than historic timescales. So some of the benefits gained by reducing public fear have almost certainly been lost and much of the remainder, by failure of the previous two siting attempts.

Geology is now, if anything, even more important because no manmade or engineering safeguards will be able to guarantee secure containment of the longer lived ILW wastes for a sufficiently lengthy duration. Indeed, it is probably doubtful whether geologic containment will succeed because of the long half-lives of some of the radionuclides. There is an argument that the principal role of any containment system, and set of multiple barriers, is to delay the return of the radionuclides for a sufficient length of time that what escapes will have only a small impact on whatever life forms are then occupying planet Earth. It should not be doubted that the longer-lived nuclides will eventually return regardless of how well the repository is sited and built. This is however at worst only going to be a problem for our descendants in 20,000 or 100,000 years time, assuming there is any humanoid life left on Earth. Nevertheless, this does create a range of moral and ethical issues about the rights of 20th–21st century peoples to leave behind radioactive rubbish tips for future civilizations to live with. It would be serious, though, if the long-lived radionuclides were to escape on historic timescales.

It is good that a multi-barrier containment system is being proposed. The solid LLW and ILW wastes are to be packaged in some suitable fashion: drums with concrete overpacking. They are to be stored in a vault within a repository which is deep underground. Movement of radioactivity away from the repository is prevented for a long, although finite, period by a suitable choice of waste form, waste container, overpack, backfilling material, vault structure and host rock. NIREX claim it should be possible to achive a very high degree of physical and chemical containment even without far field (depth from the surface and wider hydrogeological environment) protection. NIREX (1987) comment that, 'The so-called "multi-barrier approach" is designed to keep the radioactive substances away from the human environment until the process of radioactive decay makes them indistinguishable from naturally occurring material' (p.8) The health risks can be effectively minimized by disposing of the waste so that it cannot enter the food chain or drinking water and cannot be dispersed in the air. A depth of between 200 – 1,000 metres is regarded as necessary in order to provide the long far-field groundwater pathway that is required, and to reduce any chance of inadvertent future excavation of the repository or damage by some future ice age, maybe 20,000 – 30,000 years hence. All these long timescales are completely mind boggling, and very real doubts can be expressed about the competence of current technology and science to cope.

A deep repository of even the most shoddy construction should be able to achieve short-term containment of the wastes. The real question is whether this containment will survive for a sufficiently lengthy period. There are also all manner of incredible but possible mechanisms which could destroy containment: for example earth movements, a fire or bomb, another ice-age, changes in sea level, drilling, etc. NIREX can only do the best that current technology will permit. It is this technology dependency aspect that tends to result in the requirement that the waste should be retrievable at some future date should things start going wrong or a better and safer disposal option become available. At present it is difficult to imagine what could be safer. If there now has to be a radwaste dump then clearly the DCD route is about the best that can be envisaged.

NIREX (1987) identifies three concepts for a deep repository which might be sited under land or under the seabed: The three concepts are: (1) under land accessed from a land base; (2) under the seabed accessed from a land base; and (3) under the seabed accessed from an offshore structure. Caverns, tunnels or boreholes could also be used for disposal. Design studies are being carried out to establish the constraints, cost and operational safety aspects of developing one of the three basic deep repository designs. A number of these studies have already been produced considering both land and sub-seabed options. The designs for the land based repository and the sub-seabed repository accessed from the coast offer the most technologically simple solutions. In both cases the feasibility of excavations has been proven by past mining experience and is based on internationally agreed principles. Both would involve the single handling of waste, easy access and simple waste emplacement. The coastal option is, however, prone to high ground investigation costs and possible international political problems while a land based repository would suffer from NIMBY related issues. The offshore option while enjoying a certain amount of freedom from the NIMBY syndrome, has a number of disadvantages. The concept has not been proven by precedent and development costs are bound to be much higher than for the other two options in terms of research, ground investigations, repository construction and waste movements. Again as with the coastal option there could be international political ramifications. The comparative advantages and disadvantages of the three options are shown in Table 6.2.

A land-based repository could be based on man-made caverns; for example, Billingham anhydrite mine. It could also be excavated out of rock. Repository design studies all favour disposal in large caverns or tunnels since these offer the most economic utilisation of the available space. It would appear that hard rocks such as anhydrite or granite would be most suitable for large cross section

Table 6.2 Comparative advantages and disadvantages of the three siting alternatives

---

*Disposal under land*

FOR

feasibility proven by mining
simple and cheap site investigations
single handling of wastes
international consensus
easy access

AGAINST

under someone's backyard

*Disposal under seabed with land access*

FOR

feasibility proven by mining
single handling of wastes
international concensus
easy access
comparable to under land concept

AGAINST

high cost of site investigations
potentially more expensive
legally more complex

*Disposal under seabed with sea access*

FOR

avoids NIMBY problems
low ground water flows likely at depth

AGAINST

concept not proven
high cost of site investigations
additional shipping costs
additional approval problems
potentially more expensive
possible international political ramifications
legally more complex

---

Source: NIREX (1987) p.21

disposal vaults rather than mudstone or clay. In these terms a repository in one of the hard rock environments, such as the low lying hard rocks of Scotland or Anglesey and the Permain evaporites along the coast of east England, seems the most favoured option. The waste would be stacked in the vaults and backfilled with suitable grouting material (such as bentonite clay) to provide long term physical stability and establish a further chemical barrier to the migration of radionuclides.

The construction of an offshore repository with access from a mobile or fixed platform or artificial island has also been investigated. The task of constructing a shaft at sea for cavern excavation would be complex and its feasibility is not yet proven.

The geological environments considered of interest for a deep repository have been identified by NIREX, with assistance from the British Geological Survey. In looking for a site NIREX consider safety to be paramount and for this reason a deep repository is best placed in a rock formation in which groundwater movement is very slow. Chapman *et al.* (1986) and NIREX (1987) define five types of hydrogeological environment which are likely to be of interest. They are:

(1) *Hard rocks in low relief terrain.* In these areas the low relief means that there is no high land and little or no driving potential for groundwater flow. Such groundwater flow that does occur tends to be controlled by major fractures within the rock.

(2) *Small islands.* Islands which are sufficiently far from the coast have their own groundwater flow systems independent of the mainland and so are considered suitable almost regardless of their geology. Beneath the fresh/seawater interface favourable conditions occur in the form of very slow moving saline water. Should any radioactive material over long periods eventually find its way to the seabed then the dilution effect of the sea would be both massive and immediate.

(3) *Seaward dipping and offshore sediments.* In seaward dipping formations groundwater movement is expected to be very slow towards and under the coast. The driving potential for groundwater flow caused by onshore topography diminishes rapidly with distance from the coast resulting in almost zero flow conditions in sub-seabed formations. Any groundwater flow would be a very slow upwards movement caused by geothermal heating. Again, any radioactive material finding its way to the seabed would be massively diluted by the immense volume of sea water above.

(4) *Inland basinal environments.* These are deep basins of mixed sedimentary rocks containing a high proportion of impermeable formations such as mudstones and evaporites. Groundwater flow is controlled by formations of higher permeability. Flow is downward following the dip of the rocks towards the centre of the basin where near stagnant conditions have developed. The most appropriate location for a repository would be in a low permeability part of the basin limb. Any releases from a repository would first have to travel to the

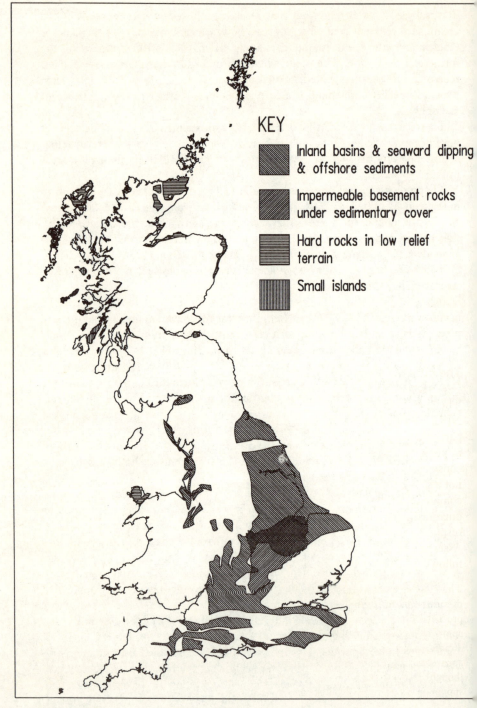

*Figure 6.6*  Areas with suitable geology for deep-cavity disposal

lower parts of the basin before moving upwards towards the surface thus creating very long radionuclide migration times.

(5) *Low permeability basement rocks under sedimentary cover*. Basement rocks (principally hard shales, mudstones, slates, volcanics etc.) occur under more recent sedimentary cover. Groundwater flows occur predominantly within the cover with little anticipated connection with the basement rocks.

The occurrence of these types of environments within the United Kingdom have been mapped by the BGS on the basis of existing borehole information, seismic surveys and written reports (Chapman *et al.*, 1986). These can be seen in Figure 6.6.

NIREX have ranked these environments in order of preference. Hard rocks in low-relief terrain, small islands and seaward dipping and offshore sediments all have simple and consequently more predictable hydrogeological environments and so are the preferred group. With these geologically suitable areas thus defined, NIREX go on to hint at ways in which areas of search will be narrowed down. Waste transport, planning considerations, environmental issues and repository design and construction are all quoted as being important factors affecting site selection. Siting arguments forwarded by NIREX in terms of transport, planning and environmental issues are very similar to those used in selecting the four second round sites in 1986.

## The current situation

The 'Way Forward' public participation exercise was designed, according to NIREX, 'to find the right combination of site and disposal concept. A major part of the discussion process is to find out which factors are the most important in people's minds' (para.5.2.13). However, the results suggest that there is no consensus on the subject of radwaste disposal. The NIREX survey is of course not a sample of any kind, nor was it a referendum, it was no more than a set of comments sent to NIREX in response to the issues raised in their *Way Forward* document. Nevertheless it is likely that most of the interested parties did respond. An analysis of the 2,526 replies by the Environmental Risk Assessment Unit at the University of East Anglia, ERAU (1988), produced the following summary of the results. They note

The most noticeable feature of the responses from these organisations is the unanimity of view from environmental groups that radioactive waste management should best proceed through some form of above-ground storage ... Further support for this approach came from a variety of other sources including Members of Parliament and Trade Union organisations ... However, there was no overall consensus ... It is apparent that scientific and advisory groups on the whole favour a deep disposal solution, though again it would be wrong to suggest that there were no dissenters from this, or that support for Nirex's disposal options was unqualified. [p.13].

A summary of the responses is given in Table 6.3.

*Table 6.3* Summary of Way Forward Response

| Source of comments | Preferences for each disposal option (%) | | | | | |
|---|---|---|---|---|---|---|
| | Deep | Under land | Accessed from land | Under Sea | Storage | Not Given |
| County/Regional Councils | 20 | 8 | 2 | 4 | 18 | 45 |
| District Councils | 13 | 5 | 2 | 3 | 9 | 64 |
| Parish/Community Councils | 1 | 2 | 0 | 0 | 9 | 84 |
| Local government associations | 14 | 14 | 4 | 9 | 9 | 47 |
| Statutory/advisory bodies | 30 | 3 | 3 | 3 | 15 | 42 |
| Private companies | 47 | 5 | 0 | 5 | 5 | 29 |
| Trade unions | 9 | 3 | 0 | 0 | 12 | 74 |
| Environment groups | 5 | 4 | 0 | 1 | 50 | 38 |
| MPs,MEPs and political parties | 11 | 6 | 5 | 6 | 26 | 42 |
| Community organisations | 1 | 3 | 0 | 1 | 13 | 79 |
| Scientific and professional groups | 5 | 26 | 10 | 0 | 10 | 36 |
| Individuals | 1 | 7 | 1 | 0 | 19 | 71 |
| Petitions | 0 | 0 | 0 | 0 | 5 | 94 |

Source: ERAU (1988) Tables 4.1 – 4.13 (values truncated, total response)

The problem with surveys is that the results can be interpreted in various ways. There is no way of knowing by how much importance to weight the responses by the various groups shown in Table 6.3. If the same data are re-analysed in the terms shown in Table 6.4, then a rather different picture emerges. The large percentage of 'not given' or don't knows is very notable; between 73 and 93 per cent of replies fell into this category. This is not a good sign. Clearly most respondents were unwilling to commit themselves to any disposal or storage option. Perhaps it reflects a lack of information, perhaps it reflects uncertainty as what is best to do, or perhaps it reflects their fears. It certainly does not give NIREX a strong mandate to do anything. With only two

*Table 6.4* Analysis of 'Way Forward' response

| Disposal option | Preferred disposal option (%) | | | |
|---|---|---|---|---|
| | All replies | All replies* | All responses† | best option(%) |
| Deep | 0 | 3 | 12 | 28 |
| Under land | 0 | 1 | 6 | 14 |
| Accessed from under land | 0 | 0 | 2 | 0 |
| Under sea | 0 | 1 | 3 | 0 |
| Storage | 5 | 19 | 74 | 57 |
| Not given | 93 | 73 | na | na |

Notes:  * excludes petition
         † excludes not given

or three exceptions all the groups have their largest response in the 'not given' category. If this category is ignored, then most respondents preferred storage over any single disposal option; 57 per cent against 28 per cent. However, this may be offering a misleading picture because there are a number of disposal options. So Table 6.5 presents the results for a comparison of disposal versus storage. Again, the 'not given' dominate but the storage versus disposal option narrows; 57 per cent against 42 per cent.

*Table 6.5* Disposal versus storage

| | Preferred disposal option (%) | | |
|---|---|---|---|
| Disposal option | All replies | All replies* | Best option(%) |
| Disposal | 0 | 5 | 42 |
| Storage | 5 | 19 | 57 |
| Not given | 93 | 73 | na |

Notes: * excludes effects of petition

Other results from the ERAU (1988) study concern the finding that the under-seabed accessed from an offshore structure had little support, even if in theory this might well be one of the safest disposal routes. The recovery of wastes was generally deemed to be important. Also of interest is the strength of concern about the potential detrimental local economic impact and blight through social stigma that seem to be associated with the public's perception of radwaste dumps, especially in areas dependent on tourism and agriculture. What NIREX will make of these responses is difficult to determine. They would certainly support either a deep sited facility, in an area which is remote, and in a location where such a development would not be considered to be blight. The apparent interest in the development of a repository under the seabed off the Cumbria coast near Sellafield, accessed by a tunnel from a point within the Sellafield site, may be one solution. However, this idea has been in existence in embryonic form for some time and test drillings were carried out and a feasibility report published as long ago as 1982. The geology and hydrogeology of the site is more complex than the three preferred environments but this site has obvious advantages because of its proximity to the principal source of waste. A similar case might also be made for an underland dump at or near Dounreay. However, the responses would also support the idea of an extended period of storage prior to disposal. Indeed if the retrievability criterion is important, then this option is the only one that will readily meet it. Unfortunately, long term storage still seems to be counter to government policy. Therefore, round three (when it starts), will almost certainly be based on a DCD proposal. If it fails, then no doubt there is only long term storage left to consider.

It is likely, therefore, that NIREX will respond by putting forward three sites for development; (1) Sellafield with a repository under the Irish Sea, (2) Dounreay, and (3) Altnabraec near Dounreay. The locational advantages in seeking sites near to the principal sources of ILW wastes are obvious. The Dounreay site is also fairly remote but would involve much higher transportation costs than Sellafield. On the other hand, the local community seems willing to act as hosts and providing there are no insurmountable technical difficulties it would make a very good site from a public acceptability point of view. It might also fit in well with the long term development of the nuclear industry. One day, in the 21st century, integrated fast breeder reactors, reprocessing and waste disposal facilities will be co-located on the same site. Dounreay and Sellafield could provide two of the possibly four major reactor parks needed to supply most of Britain's electricity in about 100 years' time. Concentration rather than dispersion of nuclear Britain would seem to make sense. A radwaste dump is a key ingredient in this long-term pattern. It is important to take such a long-term perspective because of the incremental nature of nuclear decision making that has been referred to previously.

# 7

# Searching for radwaste dump sites using geographic information systems

This chapter takes a closer look at the technology used for what is considered to be the critical preliminary screening stage of the siting process. This may appear to be merely a 'quick and dirty' trawl across Britain identifying in a crude and general manner potential sites for a subsequent more detailed examination. However, it is more important than this because of its role as a means of validating the eventual final choice by reference to the complete list of feasible locations from which it was selected. It is argued that this validation process is potentially of major significance as a device for procuring public and political acceptance. It is noted that currently NIREX appear to be using what might well be considered to be a largely obsolete manual map overlay process which could and should now be replaced by a more powerful computer based technology. Major advances in the construction of digital map data bases for Britain, decreasing computer costs and the increasing availability of Geographic Information Systems (GIS) capable of automating the map based operations suggest that NIREX now need to update and modernize their site search and screening procedures.

The last five years have witnessed major developments in practical GIS in both Britain and the United States. Most companies who deal in geographical data are now investing in GIS, for example, the utilities, British Telecom and land agencies. A number of proprietary systems are also available; for example Arc/Info (ESRI,1987). So important is this field that even the DOE had commissioned a Committee of Inquiry, chaired by Lord Chorley (DOE, 1987). Its report was published in May 1987 and recommends that effort needs to be made to expand the awareness of potential users of GIS, particularly senior managers, so that its full potential might be realized. The Chorley Report claims that GIS 'is the biggest step forward in the handling of geographic information since the invention of the map' (DOE, 1987) and stressed the importance of educating possible users in the potential offered by this form of new technology. Many of the classic GIS uses involve site searches. It would be surprising, therefore, if this new technology was not already being utilized by

NIREX in its search for an acceptable repository site as it might be expected greatly to improve several critical aspects of the radwaste dump site selection and evaluation process. Indeed, following the Chorley Report the DOE itself should now be revising its recommendations in order to provide assurances that the best available technology is being used and that the new opportunities for a more systematic and comprehensive evaluation process are being fully exploited.

GIS may appear to be little more than the replacement of a manual map based process by a computerized equivalent, but it does offer a number of important advantages and it can improve the overall credibility of the site search process. Indeed the US Nuclear Regulatory Commission (NRC) now strongly advocate the use of GIS technology in the site screening stage (NRC, 1987). The principal advantages include:

(1) it is fast, efficient and accurate;

(2) the whole of the United Kingdom land surface can be searched and explored for suitable sites;

(3) all sites are treated in an equal fashion by the same unbiased procedures;

(4) it is particularly good at handling a mixture of geological, socio-economic, environmental and policy data sources;

(5) it provides a basis for justifying the eventual siting decision which would be considerably stronger than that provided by more *ad hoc* methods;

(6) in the unavoidable impending furore, it is clearly an advantage to have an information system that can be used to demonstrate the strength of the case that is made both at the Public Inquiry and prior to it;

(7) there is today no reasonable excuse for not using these methods when trying to site such an important national facility costing several hundred million pounds;

(8) it is necessary to be able to demonstrate that all practicable steps have been taken in finding the best and most appropriate sites from a national interest perspective;

(9) GIS offers a visualization of the siting process that would be of considerable assistance in explaining how the preferred locations have been selected from the set of all potentially suitable locations;

(10) it is highly flexible allowing the effects of changes of criteria and various 'what if' questions to be speedily investigated; and

(11) experience has shown that when dealing with contentious aspects of spatial planning there is often some advantage in having a computer methodology that can be both blamed and used as defining a set of gerrymander-free rules for handling various political inputs by having an explicit (and objective – in the sense that it can be replicated) process.

The remainder of this chapter provides a demonstration of how to use a GIS as a radwaste dump screening tool and then discusses how its use can help provide a better understanding of Britain's radwaste siting problems.

## Using a GIS for the radwaste siting problem

The NIREX site search process outlined in Chapter 6 has a direct equivalent in terms of GIS operations. The procedure outlined by Beale (1985) amounts to no more than a sequence of map overlay operations. Maps outlining what are thought to be suitable hydrogeological conditions, acceptable levels of population density, conservation areas, and accessibility are overlayed to identify areas where all the specified conditions are satisfied. When this map overlay process is done by purely cartographic means, then the level of detail obtainable from even the most meticulous manual sieve-mapping must be considered, at best, inadequate given the importance of the task in hand. Most of the map patterns are complex and scale dependent, and the entire process is laborious and massively error prone particularly when combining data from different sources. The fuzzy nature of the published NIREX maps of suitable geological environments suggests that they do not have the technology to generate high quality digital maps; see for example NIREX (1987). If they used GIS, then this would not be a problem.

The lack of accuracy in Beale (1985) is especially apparent when considering geographically complex factors such as the population distribution. In mapping areas of acceptable population density, areas with more than 490 persons per square kilometre (or five persons per hectare, see Chapter 6) were excluded from the site search. Beale (1987) later reports that, 'In defining areas of search, districts or boroughs with populations of over 5 persons per hectare at the 1981 census were excluded' (p.12). A possible problem here concerns the size of these local government administrative area building blocks used by NIREX to map areas satisfying those criteria. The use of such large geographic areas obscures marked differences which occur within them and provides a national picture which is far too generalized. It will probably ignore suitable sites while including others which are not acceptable, in terms of the criteria that were used, had the calculations been performed on smaller areas. Possibly the best source of population density data is that which was calculated for the BBC Domesday interactive video disk system by Openshaw *et al* (1986) and Rhind *et al* (1986). This uses 1981 census enumeration district data to provide population count estimates for each 1km grid square. Obviously nobody can be expected to carry out 'pen and paper' sieve-mapping procedures with such detailed data (for instance there are 150,000 populated 1km grid-squares to be mapped) but the task is very simple for a computer-based GIS. These would therefore provide much greater geographic resolution and hence greater confidence in those areas selected or rejected on the grounds of population density than with any manual-map-based approach. Now it can be argued that the population criteria are not that important and that they are at most indicators which are used for purposes of broad guidance so it does not matter if they are wrongly applied! We would reject any such argument. If you have explicit criteria which are claimed to be used, then the application should be as precise as can reasonably be achieved, given the constraints of data availability.

There are also other aspects of the NIREX site search and evaluation process that must have been extremely tedious to perform by hand. Indeed, so difficult would the manual evaluation process have been, that it is suspected that NIREX would have had artificially to restrict the number of sites that could be evaluated. Beale (1987) writes: 'More detailed published geological data were also consulted and sites rated as "poor" or "marginal" were excluded from further consideration. By these means, the potential choice of sites was now down to a manageable number for yet more detailed examination of all relevant factors' (p.12). It is not quite clear as to whether the term 'sites' was a reflection of an even greater narrowing down to areas of land which were known to be under favourable ownership, or whether Beale was still thinking in terms of more generally defined areas of Britain without being constrained by site ownership. This may seem a trivial complaint but it is very important to know whether site ownership considerations had so constrained and restricted the search process that it resulted in only a very biased sample of possible locations being identified. The next stage in Beale's procedure was to weight the sites: 'A series of possible weightings was assigned to test the sensitivity of sites, given that different emphases were placed on different factors' (Beale, 1987, p.12). Ideally, these should have been calculated for the possibly thousands of area polygons that survived the preliminary screening process. In practice, this would be impossible to perform manually without greatly narrowing down the number of suitable sites. How, and if, this was done is not known. It seems likely that the weightings were applied to sites rather than areas and this restriction would bias the analysis by excluding any potentially superior sites not in favourable ownership regardless of their geological and other characteristics.

The objective here is to replicate what NIREX do, or did, in the site search part of the evaluation process. Building a GIS to automate this activity is fairly straightforward albeit somewhat labour intensive. The data capture and data management aspects are described in much greater detail in Carver (1989). Here attention is mainly restricted to the results that are obtained when the NIREX site search process is simulated using a GIS. In particular, it is of considerable interest to use GIS to replicate the round two and three siting attempts by NIREX. This involves establishing a number of data layers, each of which represents a separate set of siting criteria; see Table 7.1.

(1) The physiographic data include the various hydrogeological environments considered suitable by NIREX. The data were digitized from appropriate large-scale maps by one of the authors (see Carver, 1989). The geological areas were those used by NIREX for their site searches. Each item in Table 7.1 is stored as a separate coverage using the Arc/Info system at Newcastle University. Table 7.2 provides details on data sources and map resolution levels. Several of the key map coverages are shown in chapter 6.

(2) The environmental data consist of policy areas where there might be expected to be strong landscape and amenity objections to either a radwaste dump or to the related transportation infrastructure. Some of these data are

*Table 7.1*  Data layers

| Data type | Details |
| --- | --- |

A. *Physiographic data*

  1. Deep geological environments:
     (a) Hard rocks in low relief terrains;
     (b) Basement rocks under impermeable sedimentary cover;
     (c) Seaward dipping and offshore sediments;
     (d) Inland sedimentary basins; and
     (e) Small islands.
  2. Major aquifers
  3. Sedimentary basins on the United Kingdom continental shelf
  4. Clay geology surface outcrops

B. *Environmental data*

  5. Conservation areas:
     (a) National parks;
     (b) Areas of outstanding natural beauty (AONBs);
     (c) Heritage coasts;
     (d) National scenic areas;
     (e) Environmentally sensitive areas; and
     (f) Regional parks.
  6. Conservation sites:
     (a) Sites of special scientific interest (SSSIs);
     (b) National nature reserves;
     (c) Country parks; and
     (d) Miscellaneous sites (including ramsar sites, marine reserves and biosphere reserves).

C. *Transport and communications data*

  7. British transport networks:
     (a) Rail network; and
     (b) Road network (including motorways, A roads, B roads and unclassified).
  8. Deep water ports
  9. Shipping lanes

D. *Petrochemical resources data*

  10. Major finds:
      (a) Oil fields; and
      (b) Gas fields.
  11. Minor finds:
      (a) Oil fields; and
      (b) Gas fields.

*Table 7.1 continued*

| Data type | Details |
| --- | --- |
| | 12. Pipelines: |
| |     (a) Oil pipelines; and |
| |     (b) Gas pipelines. |
| | 13. Areas licensed for petrochemical exploration |
| E. *Demographic data* | |
| | 14. Population density |
| | 15. Built-up areas |
| F. *Boundary data (areas of interest)* | |
| | 16. UK coastline |
| | 17. UK international meridian |

stored as point references around which arbitrary circular exclusion areas were created.

(3) The transport and communications data are included as a sideline for accessibility. It is thought likely that a radwaste dump site would be located near to existing road or rail facilities. The precise distance is clearly an unknown quantity but various guesses based on previously announced sites can be made; see Table 7.3. For example, Elstow is within 0 km of a major road, 13 km of a Motorway, and 0 km of a railway line.

(4) The need to handle offshore locations explains the inclusion of shipping lanes. The location of petrochemical facilities is also important because of the possible use of a seabed disposal facility with access from a rig or artificial island. The poorer quality resolution of these data reflect the difficulty of obtaining precise locational details because of security considerations.

(5) The demographic data are from the 1981 population census and have been estimated for 1km grid squares. Various density thresholds can be used. The use of the Ordnance Survey's 1:625000 built-up area definition offers an alternative means of obtaining urban exclusion zones.

(6) The final data sets relate to the UK coastline and the international meridian. They are used only as backcloths against which other information can be displayed.

Many of the map coverages were hand digitized once appropriate map categories had been identified. It is noted that there are errors in all these data sets resulting from: (1) digitizer and map registration error, the importance of which depends on map document scale; (2) there are errors implicit in the data sources themselves; (3) there are some estimation errors in the population data; and (4) further errors created by the mixing of different data types. This implies

*Table 7.2* Data Sources and Precision

| Data item No. | Data format | Map definition | Nominal resolution | Source | Date |
|---|---|---|---|---|---|
| 1 (a-e) | vector | 1:625,000 | 10m | Chapman *et al*. | 1987 |
| 2 | vector | 1:500,000 | 10m | CEC, CWPU | 1982 |
| 3 | vector | 1:2,000,000 | 10m | HMSO, Dept of Energy | 1987 |
| 4 | vector | 1:625,000 | 10m | OS 10''geological series | 1957 |
| 5 (a-f) | vector | 1:250,000 | 10m | Countryside Commission | 1987 |
| 6 (a) | vector | 1:50,000 | 10m | Friends of the Earth | 1985 |
| 6 (a-d) | vector | 1:50,000 | 10m | Countryside Commission | 1987 |
| 7 (a) | vector | 1:250,000 | 10m | OS 1:250,000 Routemaster | 1984 |
| 7 (b) | vector* | 1:625,000 | 10m | OS 1:625,000 digital data | 1988 |
| 8 | vector | 1:50,000 | 10m | OS 1:50,000 maps BBC LVROM | 1984 |
| 9 | vector | 1:500,000 | 10m | Hydrographic Dept. charts | 1978 |
| 10 (a-b) | vector | 1:2,000,000 | 10m | HMSO, Dept of Energy | 1987 |
| 11 (a-b) | vector | 1:2,000,000 | 10m | HMSO, Dept of Energy | 1987 |
| 12 (a-b) | vector | 1:2,000,000 | 10m | HMSO, Dept of Energy | 1987 |
| 13 | vector | 1:2,000,000 | 10m | HMSO, Dept of Energy | 1987 |
| 14 | raster* | EDs | 1,000m | Openshaw and Rhind | 1986 |
| 15 | vector* | 1:625,000 | 10m | OS 1:625,000 digital data | 1988 |
| 16 | vector* | 1:625,000 | 10m | OS 1:625,000 digital data | 1988 |
| 17 | vector* | 1:2,000,000 | 10m | HMSO, Dept of Energy | 1987 |

Notes: All data coverage were hand digitized by one of the authors, Steve Carver, unless indicated by *. The OS 1:625,000 digital data base was kindly leant to Steve Carver for research purposes and the assistance of the Ordnance Survey is gratefully acknowledged.

that care must be taken when using any map overlay method, including that based on GIS. It is noted that similar error sources with much larger levels of uncertainty characterize the manual map sieving techniques that the GIS approach replaces. The principal difference is that it is now possible to take into account the errors in the GIS by performing various sensitivity analyses whereas with more traditional methods there is nothing that can be done other than pretend that they do not exist.

## Identifying the round two (1985–7) near surface sites

It is useful to start the analysis process by searching for the full set of near surface sites by replicating the manual site process that NIREX claim to have used in 1986–7 for LLW and short-lived ILW. First, areas with suitable clay

**Figure 7.1   Polygon Overlay process**

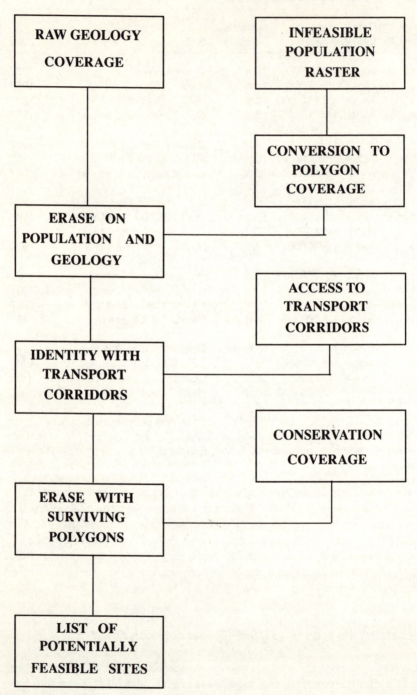

*Figure 7.1* Polygon Overlay process

*Table 7.3* Road and rail distances from previous sites

| Site name | Distance (in km) to nearest: | | |
|---|---|---|---|
| | A road | Motorway | Railway |
| Billingham | 0 | 20.0 | 0 |
| Elstow | 0 | 13.5 | 0 |
| Fulbeck | 2.4 | 46.5 | 5.8 |
| South Killingholme | 0.5 | 15.1 | 0 |
| Bradwell | 21.0 | 46.4 | 9.5 |
| Harwell | 0 | 37.5 | 4.3 |
| Sellafield | 2.0 | 82.7 | 0 |
| Dounreay | 0.7 | 381.2 | 13.0 |
| Altnabreac | 14.7 | 390.3 | 0 |

Source: BBC Domesday AIV

geologies (the London, Weald, Kimmeridge, Lias, and Oxford clays, and Kueper and Etrurian marls) were defined. The next step was to remove areas with a population density greater than five persons per hectare. NIREX lists strategic accessibility in terms of waste transportation as being important. This is interpreted as implying that the major roads (A roads and Motorways) and railway lines should be buffered, that is re-defined as three kilometre-wide corridors, and these buffer zones used to delete areas lying outside on the grounds that a 3km distance was a reasonable guess as to how far NIREX might be prepared to build new roads or rail spurs. Next conservation areas (national parks, AONBs, and heritage coastlines) are deleted. Finally, areas less than 400 hectare are dropped as being too small to consider having sufficient development potential. Figure 7.1 provides an overview of the GIS operations. The resulting feasible areas could then be assessed for accessibility to LLW and ILW waste generating sites. However, this latter process is clearly not deterministic in the same way that the previous stages were and is regarded as a post-GIS evaluation.

The results of this progressive filtering out process are shown – in terms of the effects on the size of the surviving, potentially feasible areas – in Table 7.4. The right type of geologic conditions only occur over about 15 per cent of Britain's land area. Demographic effects reduce the surviving areas to 13 per cent, the 3km rail buffer to 6 per cent, and conservation areas to 5 per cent. The inclusions of a 5km maximum distance from the coast reduces these to less than 1 per cent. The minimum site size constraint of 400 hectares can be used to identify an approximate upper limit on the number of potentially feasible sites. In plan a repository would cover about 400 hectare (four square kilometres) underground, whether deep or shallow, necessitating a ground level site directly above the tunnels and vaults of the same area since planning regulations require that these are considered part of the development. The map sequence is shown in Figure 7.2a,b,c,d. It is interesting to note that not all of the NIREX round two site suggestions lie within the final areas based on these factors.

*Table 7.4* Areas satisfying various near surface siting criteria

| Criteria | Area (in km²) | % of Britain |
|---|---|---|
| Geology (clay and marls) | 35,506.56 | 15.61 |
| +Demographic (490 per square kilometre) | 30,322.01 | 13.33 |
| +Transport (3km rail buffer) | 13,866.40 | 6.09 |
| +Conservation (NP, AONB, HC, etc) | 13,213.41 | 5.81 |
| +Size (4 square kilometres) | 12,939.49 | 5.69 |
| All above | | |
| +Coastal location (5km buffer zone) | 1,035.66 | 0.46 |

One of the benefits of a GIS approach is that it is easy to change the criteria to ask 'what if' types of question. The first of these might simply involve re-drawing the feasible sites shown in Figure 7.2d, deleting all sites further than 5km from the coastline, in order to provide a list of coastal LLW sites; see Figure 7.2e. It is noted that coastal sites are potentially 'good' locations for the dumping of bulky decommissioning wastes from power stations and nuclear submarines. It is possible that these locations may again become of interest at some future date. Additionally, the geological constraints might well be reduced or removed by a greater effort on building engineered containment, rather than by relying on natural systems. Bulk decommissioning wastes and lumps of old nuclear submarine reactor might be expected to present an easier problem and could presumably be contained purely within man-made repository designs. If this is indeed the case, then the search process could be re-run without geological constraints to produce the results shown in Figure 7.2f and the areas reported in Table 7.5. This relaxation increases the area of potentially suitable sites by a factor of four.

An additional set of 'what if' questions more relevant to LLW focuses on the demographic criteria and investigates the impact of reducing the population

*Table 7.5* Some 'what if' questions applied to near surface siting criteria

| Criteria | Area (in km²) | % of Britain |
|---|---|---|
| No geology constraint | 227,520.00 | 100.00 |
| +Demographic (490 per square kilometre) | 209,970.00 | 92.29 |
| +Transport (3km rail buffer) | 61,223.44 | 26.91 |
| +Conservation (NP, AONB, HC, etc) | 48,518.23 | 21.32 |
| +Size (4 square kilometres) | 48,193.51 | 21.18 |
| Geology (clay and marls) | 35,506.56 | 15.61 |
| +Demographic (49 per square kilometre) | 14,254.97 | 6.27 |
| +Transport (3km rail buffer) | 5,016.35 | 2.20 |
| +Conservation (NP, AONB, HC, etc) | 4,755.28 | 2.09 |
| +Size (4 square kilometres) | 4,538.72 | 1.99 |

(b) Geology, population and transport buffers

Figure 7.2  A near surface site search
(a) Geology and population

Figure 7.2  A near surface site search
(c)  ...+ conservation                    (d)  ...+ exclude small sites

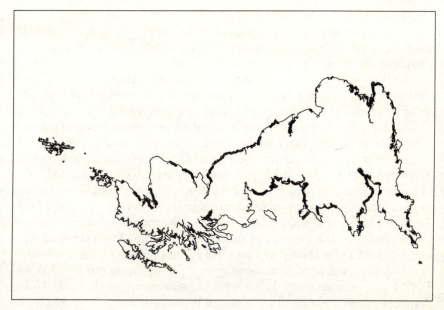

(f) Coastal sites with no constraint on geology

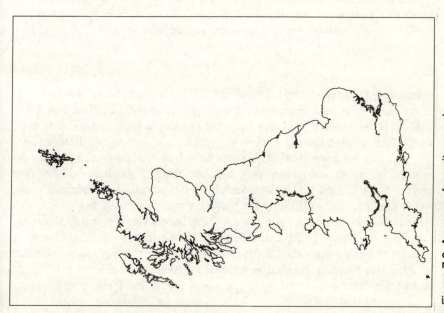

Figure 7.2 A near surface site search
(e) Sites within 5km of coast line

density limits to try and find remoter locations than are typical around the Hartlepool AGR site. So a threshold of one-tenth the previous limit is applied. Likewise, what happens if the transport access buffer is reduced in extent by assuming a rail only transport policy with the removal of the road network corridor? The results are also shown in Table 7.5. Clearly, the number of feasible sites can be made to change and the exploration of the impact of different criteria and thresholds is an important part of learning about radwaste siting in Britain. This is seen as being useful not purely for the possible benefit of NIREX, but as a means of explaining the problems and restrictions to a much wider audience – to the public, to politicians and at a public inquiry.

The other aspect of interest here is that there are seemingly surprising numbers of potentially suitable areas within which sites may be found. The question now arises as to how well the four sites that NIREX put forward perform in relation to the complete set of sites shown in Figures 7.2 and Table 7.4. It is possible to download from the GIS polygons representing all areas of greater than 400 ha in size for a more detailed evaluation. These surviving areas are rasterized and evaluated as a set of 4km grid-squares. Two sets of criteria are used. First, it is possible to compute their accessibility to various LLW and ILW wastes, and secondly, to compute populations within 3, 5 and 10 km distances. These areas can then be ranked by the population counts and by the waste accessibility scores. The best 10 per cent of sites are then mapped; see Figure 7.3.

The performance of the four NIREX sites can be explored by counting the number of potentially feasible locations that have equal or better accessibility levels. A simple accessibility to waste index is used here:

$$A_j = \Sigma_j\, W_i\, d_{ji}^{-d}$$

where $A_j$ is the accessibility score for site j, $W_i$ is the total LLW or ILW waste levels expected at each of the current waste generating sites by 2030 (see Table 2.4), $d_{ji}$ is the distance from area j to waste generating site i, and d is a distance factor (the smaller the value of d the greater the effects of distance, here $d=-2.0$ and $-0.5$ are used). This index provides a crude measure of transport costs. The results are shown in Table 7.6. None of the four NIREX sites appear to be located in highly favourable waste accessible locations. Of the other sites considered, Billingham performs only marginally better although this may well be an artefact of the accessibility index. The best waste-accessible locations are clearly Drigg and Sellafield, and perhaps Dounreay. Dounreay appears to be distance factor sensitive. It is not suggested that this accessibility analysis is sufficiently detailed to be used for real, rather it is intended as an illustration of what can and should be done, ideally using real transportation costs.

A population accessibility index can also be defined. For any site, a count is made of the number of potentially feasible locations with smaller populations within 3km, 5km, and 10km distance rings; see Table 7.6. With the exception

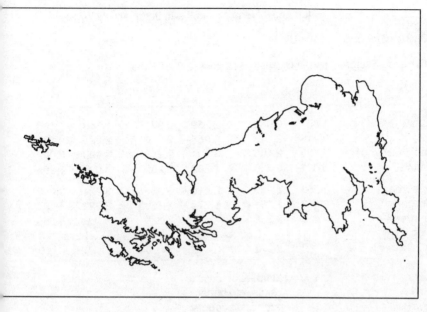

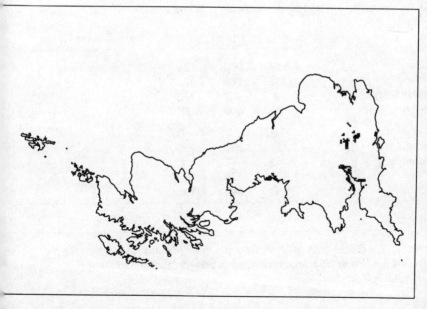

*Figure 7.3*  Best 10 per cent of sites in terms of accessibility and population
(a)  Accessibility to LLW
(b)  1km population counts

*Table 7.6* Performance of the four NIREX round two sites and other locations in terms of accessibilities and populations

Numbers of feasible sites with better accessibilities for:

| Waste category: | LLW+D | | ILW+D | | LLW+ILW+D | |
|---|---|---|---|---|---|---|
| Distance Factor: | d=−2.0 | d=−0.5 | d=−2.0 | d=−0.5 | d=−2.0 | d=−0.5 |
| **Site** | | | | | | |
| Elstow | 1,457 | 2,069 | 2,250 | 2,281 | 1,569 | 2,110 |
| Fulbeck | 2,918 | 2,146 | 2,965 | 2,078 | 2,926 | 2,135 |
| South Killingholme | 3,193 | 2,408 | 3,104 | 2,248 | 3,190 | 2,382 |
| Bradwell | 3,021 | 3,255 | 2,722 | 3,255 | 3,007 | 3,255 |
| Altnabreac | 3,251 | 3,263 | 2,416 | 3,337 | 3,224 | 3,263 |
| Billingham | 2,057 | 1,135 | 1,774 | 138 | 2,022 | 1,107 |
| Dounreay | 0 | 3,255 | 0 | 3,329 | 0 | 3,255 |
| Drigg | 1 | 0 | 0 | 0 | 1 | 0 |
| Sellafield | 0 | 0 | 0 | 0 | 0 | 0 |

Numbers of feasible sites with better population counts:

| Distance band: | 3km | 5km | 10km |
|---|---|---|---|
| **Site** | | | |
| Elstow | 2,582 | 3,000 | 2,625 |
| Fulbeck | 62 | 146 | 321 |
| South Killingholme | 2 | 451 | 755 |
| Bradwell | 482 | 590 | 75 |
| Altnabreac | 0 | 0 | 0 |
| Billingham | 3,232 | 3,235 | 3,205 |
| Dounreay | 6 | 3 | 0 |
| Drigg | 1,446 | 475 | 8 |
| Sellafield | 539 | 645 | 135 |
| Total sites | 3,263 | | |

Notes:  LLW+D:LLW plus decommissioning, commitments to 2030AD
ILW+D:ILW plus decommissioning, commitments to 2030AD
all: both above

of Elstow, all the NIREX sites perform reasonably well. The worst location is clearly Billingham. Again these analyses are meant to be suggestive rather than definitive; nevertheless, they do indicate that the 3,263 4km squares representing the feasible areas of Britain do provide a useful site pool for purposes of comparative evaluation. The number of sites is not too large for comparative study and it is time that this process became standard practice. It is important to be able to demonstrate the relative ranking of the selected sites on a national basis. As Table 7.6 clearly shows, the four NIREX sites have very different levels of performance and none is particularly impressive in terms of the national set of potentially suitable sites.

## Identifying round three (1987+) deep repository sites

The GIS-based search for deep repository sites for both LLW and ILW is basically the same as used for the round-two sites. The principal difference is that the geologically favoured areas are now different. Attention is restricted to a wider range of suitable deep geological environments and sedimentary basins on the UK continental shelf. The area of interest also extends outside the UK coastline. Other factors now become important, for instance, shipping lanes and oil reserves. The analysis is divided therefore into two parts, on land and on the seabed.

It is useful to start by searching only for under land or land accessed sites and modifying some of the previous criteria slightly to reflect experience and the different nature of a deep repository. First, areas with suitable geologies are defined. The next step is to remove areas with a population density greater than 490 persons per square kilometre, this figure being used because of the greater intrinsic safety of a deep repository and because it is a reflection of what the nuclear installations inspectorate appear to consider as acceptable for a surface sited nuclear power station. Then, areas lying greater than 3km from a railway line and 3km and 1.5km, respectively, from a motorway and main road are removed. Next conservation areas are deleted. Finally, areas less than four square kilometres in area are dropped as being too small to consider worthwhile.

The map sequence in Figures 7.4a,b,c,d shows the impact of the various data layers on the feasible areas. Again the results of the filtering out process can be shown in the terms of the size of the surviving potentially feasible areas. Table 7.7 provides a numeric summary. Potentially suitable deep geologies extend under approximately 25 per cent of Britain. Removing areas considered to be too well populated reduces this figure to 24 per cent and the effect of the road and rail corridor to 12 per cent. Removal of conservation areas further reduces this to 10 per cent and the four square kilometre site size limit to less than 10 per cent. It is interesting that much larger areas of Britain now appear to be potentially suitable, with almost 10 per cent surviving this screening process.

The effects of a limited amount of 'what if' modelling are shown in Table 7.8. In particular, the population threshold is reduced to 49, the local transport accessibility factor reduced by asssuming a rail only transport policy and the number of conservation area and site exclusions increased to the maximum.

(b) Geology, population and transport buffers

*Figure 7.4* A deep-repository site search

(a) Geology and population

*Figure 7.4* A deep-repository site search
(c) ...+ conservation

(d) ...+ exclude small sites

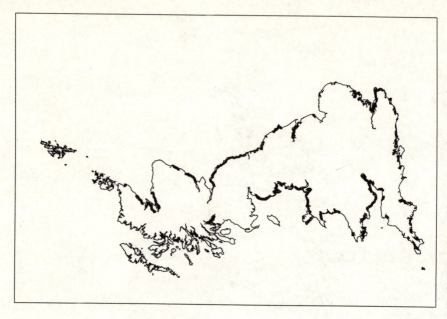

(f)  Coastal sites with no constraint on geology

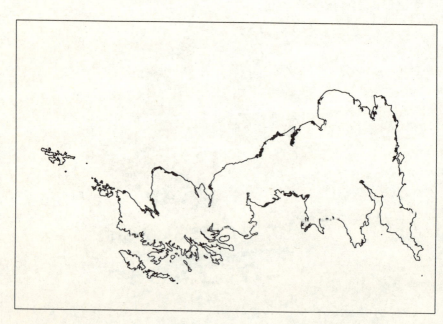

*Figure 7.4*  A deep-repository site search
(e)  Sites within 5km of coast line

*Table 7.7* Areas satisfying various deep repository siting criteria

| Criteria | Area (in km²) | % of Britain |
|---|---|---|
| Geology | 56,951.46 | 25.03 |
| +Demographic (490 per square kilometre) | 55,611.94 | 24.44 |
| +Transport (5km buffer) | 26,384.74 | 11.60 |
| +Conservation (NP, AONB, HC) | 22,581.68 | 9.93 |
| +Size (4 square kilometres) | 22,454.73 | 9.87 |
| All above | | |
| +Coastal location (5km buffer zone) | 2,553.95 | 1.12 |

The feasible areas fluctuate between 3 and 10 per cent of Britain. The surviving Figure 7.4d sites can be ranked using LLW and ILW accessibility scores and by population. The best locations from an accessibility point of view are Dounreay, Drigg, and Sellafield. This is mirrored in the population analyses, with the addition of Altnabreac. The best 10 per cent of sites are shown in Figure 7.5 and the rankings of the various sites in Table 7.9.

The search for sea accessed sites involves a similar sequence of GIS stages, although the factors considered are obviously very different. Population is no longer of importance and there are no conservation areas offshore. The location

*Table 7.8* 'What if' questions applied to deep repository siting

| Criteria | Area (in km²) | % of Britain |
|---|---|---|
| Deep geologies | 56,951.46 | 25.03 |
| +Demographic (490 per square kilometre) | 55,611.94 | 24.44 |
| +Transport (3km rail buffer) | 18,404.36 | 8.09 |
| +Conservation (NP, AONB, HC, etc.) | 16,153.80 | 7.10 |
| +Size (4 square kilometres) | 16,051.98 | 7.06 |
| +Demographic (490 per square kilometre) | 55,611.94 | 24.44 |
| +Transport (3/1.5km road & rail buffer) | 26,384.74 | 11.60 |
| +Conservation (NP, AONB, HC, etc.) | 22,581.68 | 9.93 |
| +Size (4 square kilometres) | 22,454.73 | 9.87 |
| +Demographic (49 per square kilometre) | 3,7681.44 | 16.56 |
| +Transport (3km rail buffer) | 8,890.21 | 3.91 |
| +Conservation (NP, AONB, HC, etc.) | 7,711.87 | 3.39 |
| +Size (4 square kilometres) | 7,502.04 | 3.30 |
| +Demographic (49 per square kilometre) | 37,681.44 | 16.56 |
| +Transport (3/1.5km road & rail buffer) | 13,366.15 | 5.87 |
| +Conservation (NP, AONB, HC, etc.) | 11,127.38 | 4.90 |
| +Size (4 square kilometres) | 10,787.43 | 4.74 |

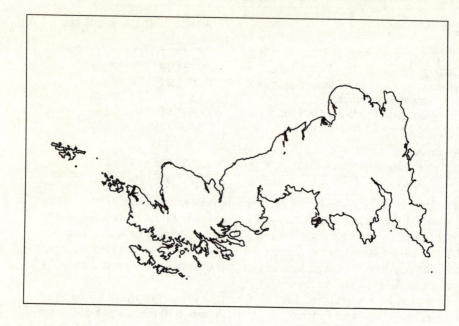

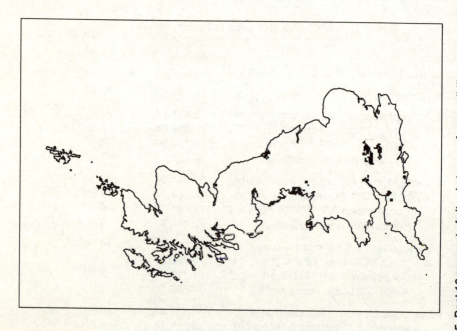

*Figure 7.5* Best 10 per cent of sites in terms of accessibility and population    (b)  1km population counts

(a)  Accessibility to LLW

*Table 7.9* Performance of various possible NIREX sites as a deep repository and other locations

| | Numbers of feasible sites with better accessibilities for: | | | | | |
|---|---|---|---|---|---|---|
| Waste Category: | LLW+D | | ILW+D | | LLW+ILW+D | |
| Distance Factor: | d=−2.0 | d=−0.5 | d=−2.0 | d=−0.5 | d=−2.0 | d=−0.5 |
| Site | | | | | | |
| Elstow | 1,872 | 3,014 | 2,870 | 3,935 | 1,994 | 3,123 |
| Fulbeck | 4,149 | 3,226 | 4,330 | 3,040 | 4,173 | 3,205 |
| South Killingholme | 4,836 | 4,115 | 4,899 | 3,857 | 4,842 | 4,088 |
| Bradwell | 4,493 | 5,259 | 3,703 | 5,279 | 4,415 | 5,279 |
| Altnabreac | 5,203 | 5,434 | 3,196 | 5,431 | 5,079 | 5,433 |
| Billingham | 241 | 1,570 | 2,309 | 1,356 | 2,510 | 1,553 |
| Dounreay | 9 | 5,300 | 6 | 5,256 | 8 | 2,984 |
| Drigg | 26 | 17 | 19 | 16 | 27 | 17 |
| Sellafield | 0 | 0 | 0 | 0 | 0 | 0 |

| | Numbers of feasible sites with better population counts: | | |
|---|---|---|---|
| Distance Band: | 3km | 5km | 10km |
| Site | | | |
| Elstow | 4,724 | 5,208 | 4,846 |
| Fulbeck | 236 | 614 | 1,086 |
| South Killingholme | 50 | 1,369 | 2,097 |
| Bradwell | 1,353 | 1,649 | 444 |
| Altnabreac | 35 | 20 | 3 |
| Billingham | 5,481 | 5,497 | 5,485 |
| Dounreay | 72 | 63 | 35 |
| Drigg | 3,141 | 1,416 | 96 |
| Sellafield | 1,489 | 1,768 | 628 |
| Total sites | 5,527 | | |

Notes: LLW+D:LLW plus decommissioning, commitments to 2030AD
ILW+D:ILW plus decommissioning, commitments to 2030AD
all: both above

of petrochemical installations is deemed to be the most restrictive factor. The overlay and filtering process is as follows. First, suitable geological conditions are established. Then areas near to shipping lanes, pipelines, and oil/gas sites/reserves are excluded. The areas which survive are shown in Figure 7.6. An alternative and speculative scenario is to consider an oil/gas field based disposal. In this case only those areas near to an existing oil or gas field might be

*Figure 7.6* Undersea sites

*Figure 7.7*  Undersea oil rig-based sites

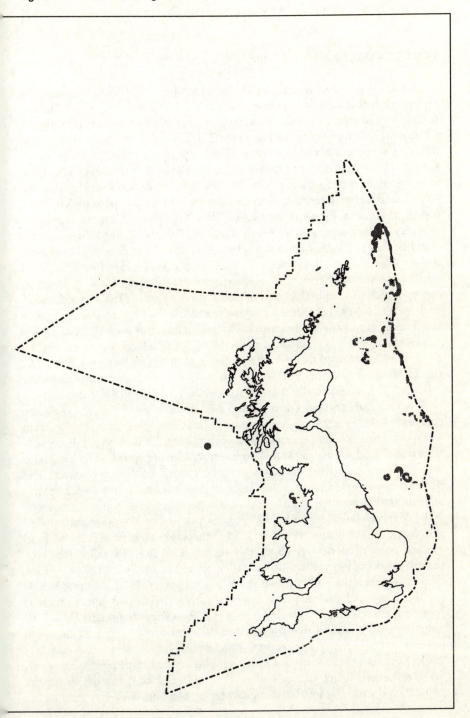

considered suitable on the basis that when construction of the repository is due to start the oil/gas field would be exhausted. Figure 7.7 defines these areas.

## What happens to NIREX's sites

A final analysis concerns checking the location of the NIREX sites against the GIS overlays that have been performed. Table 7.10 shows which of the NIREX round two and other possible sites survive this process and appear, at this stage, feasible for either a deep or shallow disposal facility and on the basis of a series of different siting criteria. All but one of the round two sites chosen by Nirex (Elstow) fails the chosen criteria for a shallow burial facility. Dounreay would also seem to fail as a deep geology site. It must however be stressed that this is a somewhat different filtering-out process than that apparently used by NIREX to identify these sites. Firstly, in the case of NIREX the process was not necessarily exclusive in that seemingly favourable sites outside the defined interest areas could be and were retained in the selection process. The South Killingholme site, for example, fails the geological criteria originally specified for shallow land burial sites in that there are no surface outcrops of clays or marls. There is, however, a considerable thickness of glacial till or boulder clays above the solid geology which were considered potentially suitable for SLB and/or ETD. For this reason (and possibly because of its availability and coastal location) the Killingholme site was maintained in the NIREX site selection process. Secondly, it must be stressed that our interpretation of the siting factors suggested by the DOE into numeric criteria may be different to their interpretation by NIREX. This is especially true of accessibility, where NIREX stressed the need to keep transport costs down and so used straight-line distances from waste sources to define a large area in the centre of Britain as being most economical in transport terms. We ignore transport costs and consider only physical access to the existing road and rail network. Our interpretation of population factors is also different despite deriving the actual numeric limits from the same source. The nuclear installations inspectorate's relaxed guidelines for siting nuclear power stations state a population limit of 100,000 persons within a five mile radius. NIREX recalculated this to be 490 persons per square kilometre and applied this figure to population data for local authorities, excluding those with an average population density above this figure. This overgeneralizes the true pattern of population distribution leading to areas whose actual population density is greater than 490 persons per kilometre square. Here the same limit is used, but population densities are derived for each 1km grid square, thereby providing a much more realistic picture of population distribution. In Table 7.10 none of the four round two sites have population densities greater than 490 persons per kilometre square, but some would fail if rigidly applying NII guidelines of population within a five mile radius. Elstow, for example has nearly 120,000 people living within five miles. Table 5.3 contains details of the population distributions within certain distances of all the sites. By changing the

*Table 7.10* Performance of sites in polygon overlay process

| | FACTORS: Surface clay geology | Suitable deep geology | Population <=490 p.km-2 | Within rail corridor | Outside conservation area | FEASIBLE AREAS: Clay sites (SLB) | Deep geology sites | No geology constraints |
|---|---|---|---|---|---|---|---|---|
| Bradwell | YES | NO | YES | NO | YES | NO | NO | YES |
| Elstow | YES | YES | YES | YES | YES | YES | YES | YES |
| Fulbeck | YES | YES | YES | NO | YES | NO | NO | YES |
| Killingholme | NO | YES | YES | YES | YES | NO | YES | YES |
| Billingham | NO | YES | YES | YES | YES | NO | YES | YES |
| Dounreay | NO | YES | YES | NO | YES | NO | NO | YES |
| Altnabreac | NO | YES | YES | YES | YES | NO | YES | YES |
| Sellafield | NO | YES | YES | YES | YES | NO | YES | YES |
| Drigg | NO | YES | YES | YES | YES | NO | YES | YES |
| Harwell | YES | NO | NO | YES | YES | NO | NO | NO |

YES = within potentially feasible area
NO = outside potentially feasible area

geological factors and using the same population, accessibility and conservation constraints, it can be seen which of these sites are potentially feasible for a deep disposal facility. Furthermore, by removing the geological constraint altogether it can be seen which sites might be suitable for the short-term surface storage of wastes. These results demonstrate what happens when a siting process is not clearly defined and when old manual procedures come up against computer technology. The general implications should be obvious.

## Using GIS

It is hoped that this use of GIS will have demonstrated some of its potential as a preliminary site search tool and have served to better inform the debate regarding the siting of radwaste dumps in Britain. There are seemingly many potentially suitable areas and the task is more or less how to find the best sites at which to locate this type of facility. If this scale of investigation can be done by the authors involving no more than about five man-years of effort then UK NIREX Ltd should be able to apply similar technology and do much better. Maybe they already do this, but if they do then we have no way of knowing because there is nothing published and no one has informed us about it. The limits to the precision of GIS are essentially those imposed by the accuracy of the original data. With suitable data high levels of accuracy can be achieved.

This GIS approach can be taken further to provide for an increasingly detailed and sophisticated site evaluation process. As has been shown, the digital 'map' of the areas surviving the polygon overlay or sieve mapping process can be reformatted in raster or grid form for further evaluation. With this approach each 4km or 400 ha raster becomes a separate 'site'. Rasterised data from relevant national data bases (for example, dominant land use, population density, accessibility to local transport networks, strategic accessibility to waste producer sites, accessibility to population centres etc) can be assigned to each of the rasters thereby utilizing the data handling capabilities of the GIS. Some of our analyses demonstrate these features. The rasters can also be scored or weighted in many different ways so that the flexibility and speed offered by GIS can be used to explore the effects of different criteria. NIREX already claim to do something along these lines, but they appear to apply this approach at a much later stage in the site search and evaluation process than might be regarded as desirable. This automatic evaluation process can be extended, so that computerized multi-criteria evaluation techniques can be used to trade-off different sets of weights and incorporate the preferences of various sets of interested or representative decision makers. This two stage approach makes the best use of the available data and would allow the GIS to become an important part of what is essentially a major decision support system. Once the most suitable areas have been defined, then attention would obviously switch to finding usable sites for further site specific investigations, secure in the knowledge that these in depth survey area(s) are 'good' places to look from a national perspective. We do not think it is adequate merely to work backwards and prove that site X is a feasible location without knowledge of its position within the total national picture. We fear that NIREX have not yet understood this basic point. With this GIS approach, the short-list of areas would reflect the conditions and constraints of Britain rather than the practical difficulties of the site search and evaluation process.

At all appropriate points in the procedure NIREX should openly communicate with the relevant local authorities and publish results as they are obtained. In this way the site selection process is kept within the sphere of public knowledge, so demonstrating that all alternatives are being considered and 'weeded out' through a rational selection procedure. Such an approach is seemingly the only way to foster public confidence in the site selection process, thereby helping to legitimize the eventual basic siting decisions. The data bases should be made widely available so that all interested parties can 'play' and 'experiment' with the many parameters of the site search process.

GIS offers, therefore, a potentially very useful debating tool to all involved in the radwaste dump siting process. While it is too crude to answer detailed site specific evaluation questions, it can nevertheless be used quite successfully to answer the types of important question that interest both public and politicians; for instance, where are the potentially feasible areas? It can be used by NIREX to justify and thus legitimize in a visibly explicit fashion their selection of sites. It can be used by government to support the political acceptability of the process on the grounds that the best available technology has been used, even if the final decisions are made for purely political rather than technical reasons. It can also be used by anti-groups to justify whatever claims and criticisms they wish to make. GIS has a considerable potential therefore as a debating tool through which all interested parties can communicate, discuss, and argue. This process is probably just as important within NIREX itself as outside. GIS can revolutionize many intra-organizational political aspects of the decision making process. It is possible for the Board itself to become closely involved in the technical processes in a painless way and with minimal risks of alienation.

All this is important because the siting process is not purely technical but requires an intense amount of public information and education. GIS has a role to play in this process. It is not strictly necessary but it does open up the preliminary site search and evaluation process in a manner which is in keeping with the new 'open door' era that is supposed to exist. The argument is expressed that, at the end of the day, most people would be prepared to accept a radwaste dump in their neighbourhood if it could clearly be seen to be the 'best' site and not just an industry convenient site from a national perspective. If NIREX or the government or the DOE merely declare by fiat that 'X' is the best site, then no one will believe them. Use a GIS, then at least there is some basis for the belief that a good or even a nearly best site has been found. However, this use cannot be purely dictatorial or deterministic but it should revolve around using GIS as a debating tool and as a common information resource.

## Conclusions

There are, however, some problems in using GIS that need to be addressed. One of these is purely technical and concerns the effects of error propagation through the sieve-mapping process. Others are institutional. In particular, the full power of the technology to identify nationally optimal sites cannot be obtained while the siting process continues to be dominated by what are essentially non-scientific methods. There are, for example, no firm numeric criteria that can be used for identifying feasible sites in relation to factors such as population or even any clear opinion that there should be some. There are no regulations as to whether conservation areas are to be regarded as absolutely infeasible areas or merely areas that it would be nice to avoid without necessarily being prohibitive. Similar comments apply to planning matters. It appears that there are no planning-related policies of sufficient importance that would preclude a radwaste dump being developed if the site was considered suitable on other grounds and in the national interest. What on earth are the DOE and the nuclear regulatory agencies playing at? The criticism is not that a wide range of material factors are being ignored, rather it is the absence of firm siting criteria than can be applied to siting factors in a mechanistic fashion during a computer-based site search. Maybe this imprecision is seen as offering NIREX some benefits by deliberately reducing the strength of the various criteria through the simple expedient of not specifying any firm or absolute numbers. Of course, this would make the DOE's task far more difficult. However, we wonder whether or not this uncertainty is actually helping or hindering NIREX. NIREX clearly have to prepare an overwhelmingly strong and carefully argued case and they might well find their onerous task easier if some of the criteria could be firmed up. This is also important because by applying these soft feasibility factors in a deterministic form will probably mean that the best areas are being missed. This is tantamount to allowing NIREX to do whatever they wish, without being restricted in any great way, then assessing the acceptability of what they do at a later date against a different set of benchmarks. This would be fine if NIREX had compulsory purchasing powers and were using GIS to perform comprehensive site searches. It is less satisfactory when NIREX are being restricted to sites in friendly ownership, when they appear to have been using obsolete methods, when very little guidance is being offered about planning and demographic constraints, and when they have essentially to guess the response of the DOE and the regulatory authorities. Within this swamp of uncertainty, the heavy hand of government can readily manipulate all aspects of the process.

There is an argument that precision at the site screening stage is irrelevant, provided that the final sites are acceptable in terms of the stated criteria. However, this is potentially the equivalent to saying that it does not matter if the calculations are 'wrong' provided that the final answer is 'right'. The debate would then return to whether a *post hoc* justification is adequate. Certainly the DOE (1985) argue that a rational site selection procedure should be used and one that works backwards might well be declared to be irrational. There is also the unwritten constraint that NIREX need to be seen to be doing their best, and the existence of possible *ad hoc*ery might well be damaging. It would, for example, make it difficult to know whether the final site was merely one of several thousand others or one of five others; or even whether it was among the 'relatively good' ones or amongst the 'relatively poor' ones. There are clearly no grounds for complacency.

There is also another potential difficulty that needs to be addressed. While details of the procedures involved in stages 1 and 3 of the site selection process are well described in both NIREX reports and related papers, the conversion of potentially feasible areas into sites for evaluation is more mystical. How specific sites are pin-pointed from the hundreds if not thousands of potentially suitable sites which appear to exist is not clear. Favourable ownership and land availability comes across as being the major factors in this process, but they need not be the only ones. Indeed, the simple fact that all four of the potential 1986–7 shallow land burial sites identified in this way are owned either by the CEGB or the MOD was enough to arouse public suspicions that the whole process had been 'gerry-mandered' to prove the suitability of one or two preferred prior locations, regardless of their true qualities as disposal sites. This is really not good enough. The failure to identify all possible candidate sites and then being able to demonstrate clearly, to the public and scientific communities, that all suitable alternatives had been carefully and properly considered without over-emphasizing favourable ownership and site availability meant that Bradwell, Elstow, Fulbeck and South Killingholme were generally held as being sub-optimal. For this reason also, GIS has an important role in establishing the linkages between the final sites and potentially feasible areas.

It is still possible to manage without GIS but the additional flexibility offered by a computer automated system is important. The power of a computer is needed to handle the multiple data layers that are involved, to allow fast experimentation, and to improve precision. Additionally, the GIS offers considerable benefits to assist decision makers to assess the alternatives. A GIS approach is also able to re-work the data quickly if the rules of the siting game are suddenly changed or if new 'what if' questions are suddenly asked. Additionally, the availability of portable and powerful microcomputer-based GIS is likely to revolutionize its impact on decision making in the near future. It is now possible to integrate senior management with the technicalities of the site search and evaluation process. GIS in board rooms and on TV are all feasible, and the media possibilities should be explored. As such it is fundamentally changing the politics of the decision making process and the means by which the results are communicated to the rest of the world. This would seem to be particularly important in the radwaste area.

Finally, there is no longer any good reason why the use of automated computer site search techniques and existing digital data bases within a GIS environment would not allow a detailed preliminary nationwide survey. It would be possible to identify all potential areas in the United Kingdom that satisfy various combinations of criteria. These criteria could either be expressed as logical restrictions or measured in a numeric fashion. The overall approach of sieve mapping of deterministic siting criteria followed by multi-attribute relative and comparative analysis of the characteristics of sites within the surviving feasible areas seems to be the most appropriate route. It is more or less what appears to be happening at present with the important exception that the comprehensiveness of the search processes, their speed and efficiency in operation, would be dramatically improved. GIS offers a quantum leap in the rigorousness by which the technology can be applied and operationalized. It offers some basis for comparative re-evaluation, for testing 'what if this condition or that criterion is changed', and for informing decision makers in a far more detailed and scientific manner than was hitherto possible. It could, by itself, offer a whole new perspective on the radwaste siting debate.

# 8

# Meeting Britain's needs for a radwaste dump

The objectives of this book are (1) to inform interested readers about the radwaste dumping issue, (2) to provide a source of facts, (3) to offer an independent review of the various siting attempts and campaigns, and (4) to identify possible areas of improvement. The perspective adopted here is that of the investigative geographer because the key questions are of a geographical nature and it is here the authors believe that the solution to the problem lies. It is taken for granted that the engineering problems can be overcome and that the principal remaining difficulties are of a socio-political nature. This is recognized by Peter McInerney (Director of NIREX) when he had this to say about the disposal of radioactive waste in 1987:

It is not an awesome task, and in fact the technical and logistical aspects of it are quite manageable. The greatest obstacle in implementing a waste disposal system is a social one, which stems from public misconceptions of the nature of radioactive material, and the risks involved in its transport and disposal. It is therefore important that the public become better informed in these areas, in order to reassure them that appropriate action is being taken on their behalf [McInernery, 1987, p.1].

The government also seems to have a similar perspective; for instance: 'The disposal of radioactive waste is necessary. The wastes exist and present facilities are either temporary or have a finite capacity. Public understanding is needed.' (DOE, 1986a, p.18). This book does not attempt to tackle the issue of how to correct McInerney's 'public misconceptions' about the reality or otherwise of the nuclear risks involved nor do we explore how the public can be made to understand! The public are not alone in having misconceptions and experiencing a lack of understanding; these problems are endemic in varying degrees to all involved in this area. Indeed, it is in this context that our repeated use of that nasty term 'radwaste dump' is deliberately provocative. It provides a better representation of what the problem is all about. Why run away and use that cosy sounding, euphamistic, and slightly misleading term 'repository'. If

177

you cannot be frank about what it is you want to build, then there are going to be many other problems downstream.

There has been a reluctance by all the actors involved in this area to face up to the severity of the problem. Indeed, it may still turn out to be incapable of any acceptable solution in the short term. No doubt there is a strong, and perhaps growing, nuclear industry feeling that the simplest solution would have been to develop the two original sites, and push them through the democratic process regardless of public and political opposition. As it happened, the sequence of Conservative governments would have been sufficient for both facilities to have been in place by now, or at least before the next general election had there been a lengthy public inquiry. However, this is Britain and not a dictatorship! The 1983 sites predated the safety assessment principles that should have logically preceded their selection. At least one of the sites was probably impossible to justify in any reasonably convincing manner. You cannot go about building radwaste dumps under towns because the geology is right and then expect public support, even if the facility is thought to be (in a theoretical sort of way) almost completely but not quite totally safe. The second round sites were clearly far better placed if it had not been for a series of basic errors: (1) the failure of the Billingham site and the attention it drew to the subject gave the opponents of any scheme a few years advance warning; (2) the decision separately to site LLW and ILW and thus implicitly create the need for two radwaste dumps, although this particular problem did not attract much attention at the time it almost certainly would have done so later; (3) the original idea for joint disposal in near surface sites would almost certainly have been unsatisfactory in a British climatic context; and (4) the Special Development Orders to avoid the planning process during site investigations looked quite sensible (early state of investigation arguments apply) but aroused the worst possible public fears and increased local authority concern. In some ways the problem in Britain is not the lack of attention to safety aspects, rather it is the perpetual tendency to ignore virtually everything else in a mad dash to operationalize an engineer's solution to the problem using 'science' as the validation tool. If it is completely or even relatively safe, then perhaps nothing else matters! Sadly, this is a fallacy. Nearly all the people who feel threatened by the development almost certainly cannot comprehend the safety arguments and would not consider the risks to be lower than the other risks of living. Indeed, not all the experts, who presumably understand the safety issues but lack faith, would agree it is safe. Honest statements of risk merely increase public fears; for instance, to say that the risk of cancer is less than 1 in a million only implies that there is a danger of cancer. Whether this 'increased cancer risk' is insignificant depends on whether you think you are one of those likely to be affected and if you believe in those who made the statement. No one wants a radwaste dump near to where he lives, regardless of whether it is completely safe or not. 'They' are the ones at risk: whatever risks exist will affect them either directly or psychologically. Fear of risk is perhaps a problem that will increase in importance in years ahead. No amount of information, free newspapers, glossy handouts, badges, audio-visual materials, exhibitions, public debate, press releases, visits, leaflets, advertisements, and public information centres will ever overcome this fundamental objection. Quite simply, if public opinion and public acceptability are important then there will not be a radwaste dump in the near future. Whatever the truth, no one wants a nuclear dump in or near their backyard; for whatever reason. Indeed, a publicly acceptable solution to the nuclear waste issue is probably also a necessary but not sufficient precondition for the long term acceptance of nuclear power. Geography has an important role to play in providing an input into the complex 'trade-offs' which look necessary in finding acceptable sites. However, whether geographical commonsense or industrial convenience or political intrigue will determine the locational aspects, is still an open question. At present, there appears to be little public support for any kind of solution, regardless of what it is or where it is located. The response to the NIREX *Way Forward* document suggests that the vast majority of those who responded prefer no solution. This complements the growing number of ordinary people who will never seemingly trust what the nuclear industry or government believes to be the truth in this area. Chernobyl has fundamentally ruined whatever slim hopes may once have existed for any approach that relies on trusting experts. Commonsense is no longer a quantity that can be applied to this area.

It is noted elsewhere that a very common misconception concerns the nature and existence of 'truth' in this area. There is no single '100 per cent correct' view that the public should or could or must adhere to, although there is a nuclear industry interpretation that would nevertheless be strongly backed by one version of modern science. However, this is not necessarily the same as the real truth and the latter is as yet unknown. In any case such arguments are academic because there are no feasible mechanisms that can be resorted to, in a democracy, to ensure that the nuclear industry's view of the world is also the public's view of the world. This book therefore makes no attempt to achieve this largely impossible task. Nor can it provide reassurance to lay readers that 'appropriate action is being taken on their behalf', action which will guarantee their safety and take into account their best interests. We do not believe that any of the participants are now in a position to demonstrate that their solutions will in fact meet this objective. Indeed, the critical comments in the various chapters might be taken to imply that appropriate action is currently not being taken. It is suggested that previous actions in this area have not been in the nation's best interest nor did they properly take into account understandable public fears and concerns. Such a conclusion is not quite what we intended, rather what we wish to imply is that there is too much uncertainty at present to know that the best and most appropriate course of action can be identified, and that once identified it will be publicly acceptable both now and for the next century or so. There are too many unknowns, too many vested interests, perhaps also too great a reluctance to 'bend' any further by giving any more ground, and a continuing failure to appreciate the need to be sympathetic to real public concerns when the steamroller option still seems to be alive. We are concerned that the best available methods and procedures are not being used, that the nuclear industry is essentially being left to regulate itself, that the apparently independent government agencies and public watch-dogs are being too apathetic and too kindly disposed towards NIREX, and that the DOE's policies are possibly inadequate for the magnitude of the problem being tackled. That department seems to have grossly underestimated the complexity of the problem and have taken a far too laid-back approach, arguing that it is a matter for NIREX and not themselves. While the reasons are understandable, this is not what might have been expected given the importance of the subject.

In making these strong statements, we also wish to express considerable sympathy with NIREX in having to solve effectively what is in reality the nation's problem without really being given the level of policy, financial, technical and practical political support needed to do so. The combined effects of public acceptability problems and political reluctance to secure all-party support while seemingly being willing to change policy direction with only modest provocation, only serve to make the NIREX task even more difficult. Are they really in charge of their own destiny? We think not, but those who are have seemingly no clear picture of the future. We want to know *why* a non-partisan and all-party approach has not been developed? If there has, what is it? Is there already in fact a consensus that disposal by what we term the 'radwaste dump approach' has strong cross-party political support, that it is suitable, and that it is likely to be broadly acceptable to the majority of the nation both now and for the indefinite future? If not, then why on earth is this option still being pursued? If there is more support for long term storage, which is also technically feasible and probably considerably cheaper, then why not switch to that? Maybe the 1982 decision to seek a permanent solution via a disposal option and the subsequent obligation by NIREX to operationalize it, should be re-assessed. There is no value in building a radwaste dump that could well be cancelled at some future date following either a change of government policy (due to the politicalization of the anti-arguments) or a change of government. How do we know that NIREX have been told to pursue the best option? The problem is that there can be no such assurance received or given. NIREX (and any successor body) are themselves hostages to fortune.

It is useful therefore to start this concluding chapter with a review of some of the key actors involved in the radwaste dumping game: the DOE, RWMAC and NIREX.

## The DOE

The DOE seem to possess a particularly clear and simple view. The government's response to the 1986 Environment Committee's review of radioactive waste was to declare 'Each society has the responsibility of managing the waste it creates. It is an abrogation of that responsibility to leave the problem for a future generation to resolve. The technology and the science exist. The necessary facilities should be provided. Even so, this will only be possible if the public understand and accept the case being made.' (DOE, 1986a, p.6). This is an accurate but somewhat arrogant view that suggests more than a little impatience. It was accepted while the opposition was merely a vocal fringe minority for whom anarchy and anti-nukes are a way of life. However, the opposition is no longer a small eccentric minority but probably now a majority of the public at large in the 'affected' areas. The events since Billingham and Elstow have suggested a very different scale for the public acceptance problem. The DOE message appears merely to be that here is the solution so why not accept it! The problem is that many people do not seem to like the favoured solution. There is indeed an alternative in the form of long term storage, but this has been dismissed as implying an abrogation of responsibility. The validity of this argument is fairly weak. It would also be an abrogation of responsibility and democracy to impose a solution that no one likes and which future generations may have to reverse. Nothing in the radwaste dumping business is easy.

The DOE's BEPO statements provide some justification for a dumping solution. But was it the result of the best available scientific advice, or was it, as seems so common in the nuclear area, the result of actions by the 'friends' of the nuclear industry? Is it possible to obtain independent nuclear-related advice on anything in the United Kingdom using experts who tend to have originated from within the nuclear industry or who were trained by it? What proportion of these 'independent' advisers are really independent in thought, word, and deed – or is this impossible in the British context? Certainly, some of the statements by the so-called 'public watch-dogs' smack of a strong lack of independence and may even be a hidden mouthpiece for the nuclear industry itself. There is nothing obscene or tremendously bad about this, if these organizations are also able to identify independent commonsense-based alternative solutions of their own rather than seemingly echo already entrenched hardline nuclear industry views.

The quick dismissal of what might be considered proper siting principles by both the DOE (1985) and then later NIREX is surprising (see previous chapters for an extended discussion). The DOE (1985) itself reports the opinions received after the publication of their 1983 draft siting principles. It notes that, 'The overall aim should be the "best practicable environmental option". The explanation of why the optimal site need not be chosen was unclear. (p.21). The grounds for this change in DOE emphasis between the 1983 draft and 1985 final version is seemingly still a mystery to both ourselves and also to the DOE. The DOE are after all supposed to be independent of NIREX but here, presumably on the advice it received from NIREX or their advisers, it is implying that comprehensive searches for optimal and nearly optimal locations are unnecessary. Why? Is it because the process is really considered unnecessary or is it impracticable given manual map-based analysis technology? Yet in many ways this is the key aspect of the entire radwaste dump siting process. How else can the government and public be confident that

NIREX have done their best? The comparative evaluation of at least three alternative sites is no substitute for a fully comprehensive evaluation process. What are they frightened of? Planners and decision makers presumably learned long ago how to select a handful of alternatives in such a manner that they merely emphasize the virtues of the preferred location. If the comparison of alternatives is important as an aid to legitimation, then many more than three or five sites are needed if the problem of how the alternatives were chosen is to be avoided. On the other hand, if the target site is known in advance of any search being initiated then the need to establish its relative performance in relation to some unrealistic optimal site would only serve to confuse the process, or so it might be claimed. The counter argument is simply that the best way of demonstrating the acceptability of a pre-defined or any other favoured site, is by putting its properties in the broader context of *all* the alternatives that exist. This provides an overview of the whole siting universe and avoids the dangers of bias and claims of cheating that may otherwise arise.

It still seems astonishing that despite having explicit and seemingly fixed criteria, those involved see no need to be constrained by the initial site searches. Sites which fail the declared 'fixed' criteria can still be judged acceptable if they possess other useful properties. Clearly it is not a scientific process when the criteria are not fixed a priori and can be modified or ignored during the site search and evaluation process for undefined ulterior reasons. Add a high degree of secrecy or 'normal commercial confidentiality' and there is no certainty about any aspect or any claim that is made. Given this flexibility, no wonder McEwen and Balch (1987) show little enthusiasm in searching for the 'best possible' site(s). It is true that there are some basic misconceptions here. It is probably impossible to find an optimal site with respect to all the criteria that could be applied, but this is no justification for not trying. The term 'optimal' also carries with it a presumption of precision which would appear to be inappropriate in this context where there is a broad mix of data with a wide spectrum of error levels. It also requires a full implicit or explicit evaluation of all the possible locations in terms of the same set of criteria and this task might be considered impossible. Indeed, siting major nuclear facilities was never a matter of simply looking for a single optimal location. Rather it is all about looking for good locations from among the list of nearly optimal or 'good' alternatives. However, the notion of optimality is important mainly because it can be used as a benchmark against which other inferior locations can be compared. Only then is it possible to put the preferred locations into the wider national context and thus apply the full strength of the national interest argument. We have repeatedly argued the importance of this type of justification on technical grounds and because it would be a powerful prop in support of any attempt to carry a greater share of public and political opinion. It is also the least that might be expected in planning a facility likely to cost several hundred million pounds.

If NIREX really want to persuade people that they are doing their best to find the best location with the best standards of safety, then they need to switch from a state of being happy with feasible sites to a search for nearly optimal locations. Whether the DOE consider this essential or not is to some extent irrelevant; the key criterion is whether or not NIREX is able to present the strongest possible case and the onus to do so rests with NIREX rather than the DOE. However, no doubt the DOE could help more by clarifying the siting criteria and by setting explicit rules within which the subsequent siting debate can take place. Vagueness may be an advantage in that it gives NIREX as much freedom as possible, but it also provides no real framework for the wider debate. The DOE should also be urging greater attention to all aspects of the site search and evaluation process, including the adoption of the best available site evaluation technology. Anything less and they are failing in the spirit of the uniquely important responsibility entrusted to them. It is surely the DOE's responsibility to shoulder more of this burden.


## RWMAC

The Radioactive Waste Management Advisory Committee (RWMAC) was created to provide advice to the government on major issues relating to the development and implementation of overall policy for the management of civil radioactive waste. The committee is made up predominently of independent members with relevant backgrounds or expertise. It often appears to favour a hard-line pro-nuclear stance. For example, in their comments on the NIREX *Way Forward* document, they report

RWMAC have been assured that NIREX is prepared to bear any construction cost that is required to meet safety and environmental needs. We applaud the intention but would not want to see wholly unreasonable and unnecessary demands made on NIREX for further expenditure beyond these requirements; there are other areas of radiological protection which could be far more profitably pursued with such additional expenditure [RWMAC, 1988a, p.4].

What possessed RWMAC to make such a statement? The money saved by NIREX will not be available for any other kind of radiological protection. This same argument has been repeatedly used by the nuclear industry to justify not spending money on safety matters. It is at best a misuse of cost benefit analysis that has no relevance. Surely, NIREX are entirely justified in spending whatever monies are necessary to guarantee public safety but this is of no concern for RWMAC! The nature of a radwaste dump is such that here, more than anywhere else, in the nuclear state, there can actually be some guarantee of safety. Instead of saying that it will cost k-million pounds to reduce the cancer risk from 1 in 1 million to 1 in 10 million and then arguing that the investment is not worthwhile, they should be declaring that there is no cancer risk to the public associated with the radwaste dump. If this cast-iron assurance cannot be given then something is potentially horribly wrong. If the risks are non-zero then people will feel threatened. With nuclear power reactors, it is quite reasonable to quantify the risks involved and to trade them off against the possible value of the benefits of nuclear electricity. A certain degree of danger is unavoidable because nuclear reactors are highly complex industrial systems; public safety depends on engineering design, active systems and operator skills so it is hardly surprising that there are unlikely but feasible accident scenarios that could have a devastating effect. However, with a radwaste dump, the design should be such that the risks are zero. There are no live systems that have to continue to function without error in order to protect people. Even if the real radwaste dump risks are not zero, and nothing created by man has zero risk, nevertheless, NIREX should be prepared to underwrite a public zero-risk guarantee. If they believe their own safety assessments, then this will never cost anything and it may well make many people 'feel' much happier.

RWMAC also comment on the subject of population density. They note, 'This is an important consideration but we see it more as an environmental issue than in relation to safety. We consider it essential that any facility is built to absolute safety standards irrespective of the size and disposition of any neighbouring community. For nearby populations the most important additional consideration must be the aesthetic impact, and the physical and social consequences of traffic at the site'. (RWMAC, 1988a, p.4). This is not as unreasonable a statement as it may appear. Obviously, any repository has to be safe but it is the term 'absolute safety standards' that implies less than perfect safety and it becomes doubtful whether any local community would accept this sort of conditional assurance. Why should they shoulder additional risks, albeit small, for the nation as a whole, when all they have to look forward to is a century of blight, possible health risks, and repeated media campaigns that can serve only to heighten local fears and concerns? Additionally, the strength of the NIMBY factor is certainly population-related. It is again 'strange' that RWMAC should be so 'blind' to these broader siting aspects. Has the Billingham fiasco not taught them anything? Is it possible that the entire Committee has been seduced by the nuclear industry arguments and are prepared to ignore public acceptability issues solely because they are men of science? Certainly, it seems that here the engineer's view of the world reigns supreme. Political naïvety also appears to be a strong element. Perhaps RWMAC are to blame for some of the early NIREX mistakes. Alternatively, maybe RWMAC's enthusiasm to solve the nation's radwaste dump problem in a speedy fashion is to blame; their haste might also reflect strategic skills and foresight that no one else appears to possess. Perhaps they alone properly appreciate the urgency of the task.

However, we would argue that the public and mass media will never view population factors as irrelevant or that landscaping is a suitable device by which to overcome NIMBY-related effects. A more considered approach would be to design a repository which is judged to be safe and then to site it remotely so that NIMBY effects can also be minimized. No one is going to believe that safety standards can be maintained for ever so why pretend they can be? A remote site would at least provide an additional, engineering and geology independent safety mechanism. Or is it that the 'best' sites only occur in non-remote areas?

RWMAC also made some statements about site optimality. They write, 'The search for a suitable site must not go on indefinitely; there is little advantage in searching for a "Rolls Royce" solution if NIREX find the answer in a "Jaguar" ' (RWMAC, 1988a, p.6). The problem is how can you be certain that it is not a rusty 'Lada'? Also, why the haste? Obviously no siting process can continue indefinitely but should it not continue until there is some assurance that the best sites have been found? A 'Rolls-Royce' solution would now appear to be mandatory. Moreover, why should RWMAC be bothered about cost and time factors, neither of which affect themselves or the government whom they advise? The Committee state that they would support a pragmatic approach to site selection provided corners are not cut. They seem to be keen to expedite the entire process in case there is another change in policy, but don't they realize that the lead times are sufficiently great to require the support of two or three, or more, governments before a facility is likely to be fully operational and thus immune from cancellation on political grounds? Unlike nuclear power stations, half-finished radwaste dumps are eminently suitable canditates for cancellation. RWMAC also advocate plans to be made in case NIREX is unable to pursue its current objectives, leading to an undefined period of prolonged storage. Once again, for reasons which are not apparent, the Committee seem to be more concerned about the waste disposal problem than any of the other agencies involved. Perhaps they alone are doing their job properly?

## NIREX

NIREX have shown considerable fortitude in facing a very difficult task. They should seriously consider whether GIS can help them cope, if they still do not use this technology. GIS will be useful not so much now with the site identification task more or less complete, but with the site justification phase and with future site searches for additional facilities. Indeed it must now be difficult to justify not using GIS, although its relatively late availability in Britain may well account for NIREX's apparent reluctance to use what might still appear to be a research rather than an operational tool. A more serious restriction is that before GIS can be fully and properly used to its full potential, it will be necessary to flush away some of the siting attitudes and folklore surviving from the past. The nuclear industry had a narrow escape at the Sizewell PWR Public Inquiry. The inspector asked but did not pursue the question as to whether the environmental devastation at Sizewell could have been avoided by developing a different site. At Sizewell the debate was narrowly focused on the acceptability of a single site taken in isolation, and there was apparently no great need to compare Sizewell with the full set of possible alternative locations. This could presumably be justified at that time by the difficulties of coping with a wider site search and evaluation exercise using manual methods and by the inspector referring this matter back to government whereupon it was 'lost'. However, the former excuse no longer applies and the importance of the wider evaluation process is clearly now of greater significance. This implies that NIREX will need to consider carefully how they can justify, in public, their site selection process and, perhaps, even illustrate the problems they faced and the nature of the solutions they advocate. It is also possible that a radwaste dump Public Inquiry will not repeat the Sizewell mistake by failing to put sufficient emphasis on these broader aspects of site selection and validation. In particular, it will almost certainly want to know how well the proposed site performs in relation to a much larger number of alternatives; if only because the benefits from a radwaste dump are considerably less compelling than from a nuclear power station. If NIREX cannot provide the answers then no doubt some of the anti- groups will do so, especially if they start to use GIS. It can also be argued, that NIREX are in an inherently far weaker position than the statutory undertakers in a similar situation and they will have to 'fight' much harder to win the key battle for scientific credibility and political support, even if there were naturally a strong presumption in their favour.

However, GIS does have to be used carefully. American experience suggests that the media effect of showing maps of potentially feasible sites can be devastating. Possible or potentially suitable areas become translated by the media into 'target' areas, with an unjustified and major change of emphasis. It also means that areas of Britain flagged as being 'excluded' cannot easily be 'included' again. Publicity in this area is a one-way process and, doubtlessly all publicity is harmful to whatever case is being prepared. Radwaste siting maps are the ultimate media horror story! Knowledge of the existence of potentially large numbers of possibly suitable radwaste dump areas does nothing to reduce public fears and neuroses. All the people living in possible target areas may well be convinced that it is their town and their neighbourhood that is being targeted. Couple this with non-committal statements and you create major media 'shock-horror' headlines that can be repeated in every local newspaper in the country. Seemingly, this most critical phase has now been passed in the United Kingdom. NIREX have already published (in their own newspaper, *PlainTalk*) various maps so it hardly makes much sense not to continue the process as a means of trying to achieve a better informed public. The conditional nature of the displays need to be demonstrated by offering a number of different sets of results, in order to try and avoid the danger that areas once excluded cannot later be included again. It is useful to be able to explain how a particular set of potentially feasible areas were chosen, and what the effects of different criteria would be. It is also important that there is a degree of trusting the public rather than paranoia, and a willingness to be more open instead of secretive. It would now appear that NIREX is heading in this direction. Whether this process continues is a matter of speculation.

**The real 'Way Forward'**

It is easy to be critical and far more difficult to make constructive suggestions. It is hoped that some of our comments will be viewed in a positive light. In their *Way Forward* document NIREX seem to have identified what is probably the optimal strategy. The key criteria are not geology or safety, both can be assumed if the 'homework' is done properly (as it will be), but public acceptability, both now and in the future. Site location is the key component in this process. It is important to find one or more sites which meet the following constraints: (1) it satisfies the DOE safety assessments; (2) it is in an area where there is likely to be a high level of public acceptability; (3) the site can be shown to be suitable on the basis of siting factors; (4) the site is one of the best that could be identified in terms of the nation as a whole; and (5) the site can be developed. These requirements can probably be met because there is a sufficiently large area of Britain that appears to be potentially suitable for radwaste dumps. For instance, Table 8.1 shows for each county the areas of

*Table 8.1* Distribution of feasible areas by county

| County: | Percentage total feasible area by county: | | |
|---|---|---|---|
| | No Geological constraints | Round 2 areas (SLB facility) | Round 3 areas (Deep facility) |
| km²: | 45,644.30 | 12939.49 | 22454.73 |
| Percent of total land area: | 20.06 | 5.69 | 9.87 |
| 1 London | 0.00 | 0.00 | 0.00 |
| 2 Greater London | 0.20 | 0.30 | 0.09 |
| 3 Greater Manchester | 0.44 | 0.42 | 0.21 |
| 4 Merseyside | 0.03 | 0.17 | 0.42 |
| 5 South Yorkshire | 1.47 | 0.06 | 1.07 |
| 6 Tyne and Wear | 0.16 | 0.00 | 0.00 |
| 7 West Midlands | 0.54 | 1.21 | 0.66 |
| 8 West Yorkshire | 1.24 | 0.00 | 0.00 |
| 9 Avon | 0.83 | 1.58 | 0.10 |
| 10 Bedfordshire | 0.78 | 1.69 | 1.06 |
| 11 Berkshire | 0.61 | 1.18 | 1.04 |
| 12 Buckinghamshire | 1.40 | 3.51 | 2.64 |
| 13 Cambridgeshire | 2.59 | 4.57 | 4.84 |
| 14 Cheshire | 2.50 | 5.22 | 1.32 |
| 15 Cleveland | 0.49 | 0.38 | 1.34 |
| 16 Cornwall | 2.14 | 0.00 | 0.00 |
| 17 Cumbria | 3.73 | 0.60 | 0.98 |
| 18 Derbyshire | 1.74 | 2.05 | 0.36 |
| 19 Devon | 0.31 | 0.70 | 1.41 |
| 20 Dorset | 0.63 | 0.48 | 0.83 |
| 21 Durham | 1.05 | 0.00 | 0.25 |
| 22 East Sussex | 0.59 | 0.59 | 1.39 |

| County: | Percentage total feasible area by county: | | |
| --- | --- | --- | --- |
| | No Geological constraints | Round 2 areas (SLB facility) | Round 3 areas (Deep facility) |
| 23 Essex | 1.65 | 6.94 | 0.00 |
| 24 Gloucester | 1.20 | 2.44 | 1.66 |
| 25 Hampshire | 2.16 | 1.23 | 2.39 |
| 26 Hereford and Worcester | 1.79 | 2.94 | 1.97 |
| 27 Hertfordshire | 0.97 | 1.05 | 0.01 |
| 28 Humberside | 2.50 | 3.13 | 7.51 |
| 29 Isle of Wight | 0.00 | 0.00 | 0.00 |
| 30 Kent | 2.78 | 3.44 | 2.01 |
| 31 Lancashire | 2.26 | 1.03 | 1.21 |
| 32 Leicestershire | 1.98 | 4.29 | 4.58 |
| 33 Lincolnshire | 4.46 | 0.04 | 12.52 |
| 34 Norfolk | 3.37 | 1.77 | 5.92 |
| 35 Northamptonshire | 1.60 | 0.90 | 4.98 |
| 36 Northumberland | 1.64 | 0.00 | 0.00 |
| 37 North Yorkshire | 4.33 | 3.21 | 6.69 |
| 38 Nottinghamshire | 2.49 | 4.79 | 3.01 |
| 39 Oxfordshire | 1.51 | 2.66 | 2.90 |
| 40 Shropshire | 2.19 | 0.88 | 0.48 |
| 41 Somerset | 2.19 | 4.28 | 3.56 |
| 42 Staffordshire | 2.28 | 3.34 | 0.71 |
| 43 Suffolk | 1.39 | 0.98 | 0.01 |
| 44 Surrey | 1.16 | 0.53 | 0.45 |
| 45 Warwick | 1.62 | 4.82 | 3.84 |
| 46 West Sussex | 0.70 | 1.20 | 0.26 |
| 47 Wiltshire | 1.92 | 3.53 | 3.40 |
| 48 Clwyd | 0.47 | 0.00 | 0.13 |
| 49 Dyfed | 2.78 | 0.00 | 0.00 |
| 50 Gwent | 1.33 | 0.76 | 0.00 |
| 51 Gwynned | 1.34 | 0.00 | 1.21 |
| 52 Mid Glamorgan | 1.37 | 0.58 | 0.00 |
| 53 Powys | 1.76 | 0.00 | 0.00 |
| 54 South Glamorgan | 0.37 | 0.92 | 0.00 |
| 55 West Glamorgan | 0.63 | 0.05 | 0.00 |
| 56 Borders | 0.25 | 0.00 | 0.00 |
| 57 Central | 0.18 | 0.00 | 0.00 |
| 58 Dumfries and Galloway | 0.00 | 0.00 | 0.09 |
| 59 Fife | 0.00 | 0.00 | 0.00 |
| 60 Grampian | 2.56 | 0.00 | 1.51 |
| 61 Highland | 6.86 | 0.27 | 2.47 |
| 62 Lothian | 0.28 | 0.00 | 0.26 |
| 63 Strathclyde | 1.65 | 0.00 | 0.01 |
| 64 Tayside | 1.45 | 0.00 | 0.00 |
| 65 Islands | 0.00 | 0.00 | 0.00 |

potentially feasible sites for the three types of radwaste dumps examined in Chapter 7; a surface store, shallow land burial facility and a deep disposal cavity. It is difficult to accept that 99 per cent or more of these large areas of potential sites are all going to be unsuitable for radwaste dumps on other grounds, or that the land ownership question really need be so constraining.

In seeking to meet the list of basic requirements it is noted that there is an argument for considering the operating economics and the cost of the development to be largely incidental. It would be silly to pick a location for a facility that will probably have an operational life span of 100 years based on minimizing transport costs as they exist in the 1980s. There has to be a radwaste dump and there is clearly a monopoly situation, so it is very much a seller's market and the construction of a national facility at virtually any cost will soon become a major commercial concern. Perhaps the only critical cost factor is the expected operational life over which the capital cost is to be discounted. Provided this is sufficiently lengthy, then there cannot be any real difficulties in seeking an expensive 'Rolls-Royce' solution. It has been argued elsewhere that maybe the cost of buying public acceptability is to seek and pay for the development of sites which have as few people as possible living near them and then construct the safest possible facility. Remote siting is still considered feasible, even in the United Kingdom. It is particularly applicable to long term surface storage facilities which are essentially 'geology independent' and also to other disposal options which are not. Table 7.5 describes the effect of a search for surface storage sites that are geology independent (see Figure 8.1). It is noted that over 21 per cent of the country would appear suitable, assuming all the other restrictions still apply. An analysis of the waste accessibility levels and population rankings is given in Table 8.2. There is probably no reason to switch from the currently favoured locations; note in particular the superior population rankings of Altnabreac, Dounreay, Sellafield and Drigg. In terms of accessibilities, it is Drigg and Sellafield, and to a lesser extent Billingham, that triumph. Figure 8.2 shows the locations of the top 10 per cent ranked sites, by accessibility and population. In terms of the broader political debate the distribution of potentially suitable sites are in fact fairly widely distributed around Britain. This aspect tends to have been overlooked.

However, it should also be recognized that solving the nation's radwaste dump problem will earn no knighthoods and cause only continuing problems. NIREX is seemingly the politician's whipping boy, the fall guy who is paid for by non-government sources and who is the ultimate excuse whenever policy deficiencies surface. This situation might be expected to last only for a relatively short period during the facility construction period, after which NIREX will have proved itself and become a major economic asset. On the other hand, if NIREX get it wrong, then the cost of the subsequent recovery operations might well be enormous – if of course anyone was to notice that problems exist. One of the attractive features about deep disposal is that problem existence should take thousands of years to be recognized; with shallow burial facilities time periods of a few decades might be sufficient. It is probably only a matter of a few more decades before something will have to be done about Drigg.

*Figure 8.1* Potentially suitable surface storage sites

*Table 8.2* Performance of various possible NIREX sites as a surface store

| | Numbers of feasible sites with better accessibilities for: | | | | | |
|---|---|---|---|---|---|---|
| Waste category: | LLW+D | | ILW+D | | LLW+ILW+D | |
| Distance factor: | d=−2.0 | d=−0.5 | d=−2.0 | d=−0.5 | d=−2.0 | d=−0.5 |
| **Site** | | | | | | |
| Elstow | 2,646 | 3,692 | 4,133 | 4,254 | 2,862 | 3,668 |
| Fulbeck | 5,273 | 3,754 | 5,425 | 3,549 | 5,296 | 3,732 |
| South Killingholme | 5.806 | 4,524 | 5,826 | 4,191 | 5,811 | 4,484 |
| Bradwell | 5,528 | 6,031 | 4,986 | 6,054 | 5,491 | 6,052 |
| Altnabreac | 5,994 | 6,216 | 4,471 | 6,212 | 5,919 | 6,215 |
| Billingham | 3,709 | 2,037 | 3,298 | 1,816 | 3,664 | 2,021 |
| Dounreay | 8 | 6,072 | 6 | 6,029 | 7 | 6,072 |
| Drigg | 28 | 17 | 19 | 16 | 28 | 17 |
| Sellafield | 0 | 0 | 0 | 0 | 0 | 0 |

| | Numbers of feasible sites with better population counts: | | |
|---|---|---|---|
| Distance band: | 3km | 5km | 10km |
| **Site** | | | |
| Elstow | 4,936 | 5,717 | 4,976 |
| Fulbeck | 213 | 464 | 830 |
| South Killingholme | 57 | 1,035 | 1,726 |
| Bradwell | 1,101 | 1,287 | 332 |
| Altnabreac | 33 | 20 | 2 |
| Billingham | 6,227 | 6,242 | 6,195 |
| Dounreay | 78 | 65 | 38 |
| Drigg | 2,851 | 1,075 | 104 |
| Sellafield | 1,216 | 1,391 | 453 |
| Total sites | 6,300 | | |

Notes: LLW+D:LLW plus decommissioning, commitments to 2030 AD
ILW+D:ILW plus decommissioning, commitments to 2030 AD
all: both above

From the public point of view, the optimal strategy is as follows:

(1) insist on the development of a 'Rolls Royce' solution since it will cost the taxpayer nothing;
(2) argue for the highest possible levels of safety because the expert

assessments are largely theoretical rather than empirical facts and are prone to unknown levels of error and uncertainty;

(3) in case all else fails insist that only a very remote site is used and this reduces the NIMBY phenomenon;

(4) critically examine all statements and actions by government departments and agencies in this area because of the possible hidden influence of the nuclear industry; and

(5) offer the host community some positive financial inducements to ensure acceptance and establish a cadre of happy workers who will guarantee that safety standards are maintained. The latter is very important. Why not pay a community to host wastes? The idea is alien to British life but it is after all a market-based solution and radwaste is a very exceptional commodity.

Since the eventual justification for the development is the national interest, it is important that the strongest and best possible scientific case is made. There is no excuse for skimping on the search and evaluation process. The nation needs to know that site 'X' has be used for a radwaste dump, not because it is owned by the nuclear industry, not merely because the geology is right, not only because it is convenient, but because it can be shown to be one of the best sites in the country as a whole. Public access to basic site search and evaluatory GIS would seem a logical development to try and provide the level of case justification that may well be required.

Another important factor here concerns the use of limited duration surface storage as a substitute for immediate disposal. The results of the 'Way Forward' public response exercise only really showed one result of major significance: most of the replies specified no particular solution. Is it possible that most of the respondents were totally indifferent to the various options and would quite happily accept any of them? The answer is probably 'no'. They either did not like any of them, or else they did not understand what the options really were all about. Maybe it is like asking the prisoner who is about to be executed which method of death he would prefer: firing squad, hanging, electrocution – or not stated! If the vast majority of the population really do not like any particular disposal or storage option, probably because approval might draw attention to their own location, then long term storage could be the answer. The DOE's BPEO ruled out storage unless retrievability of the wastes was considered important. However, they did not include possibly greater public acceptance as relevant probably because the 'benefits' cannot be quantified. People seem to fear disposal because of its permanency, the longevity of the risks, and proximity to themselves. Maybe they also do not trust what the experts tell them and maybe they also do not understand the detailed arguments. Other anti-nuclear experts may well contradict the pro-nuclear experts and start a debate as to who is 'right'. Well maybe they are both 'wrong' but at present no one knows who or what will be considered right in 50 or 100 years time. So why not 'buy time' by using the available technology of intermediate storage. Plan for 50 – 100 years by which time the intensity of the nuclear committment will

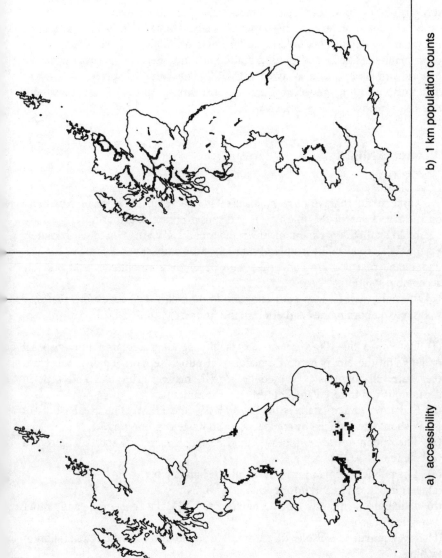

a) accessibility

b) 1 km population counts

*Figure 8.2* Best 10 per cent of surface storage sites

have clarified many of the public acceptability issues. Why rush and face failure now when the whole business can be left to our children's children to sort out for us? There is nothing immoral or unethical in that and this is not really an abrogation of our responsibilities for waste management. If we rush ahead with ineffective disposal options, then that also is an even graver abrogation of our responsibilities to future generations. Indeed, if the current deep disposal proposal fails to materialize then this long term storage option will have come into being by default. It would be better however, to plan for it in a honest and open manner. There is after all a real opportunity here to get away from the historic problems and also avoid NIMBY. The latter is purely a matter of manipulating siting geography and surface storage provides the maximum degree of freedom to do precisely this.

## A Letter to NIREX

Dear Sirs,
We understand that you are still active in the radioactive waste repository business and are in the market for a disposal option. Furthermore, we have been told by the Department of Environment and RWMAC that it is absolutely essential that this facility is up and running at the earliest possible date. We also understand that you are concerned with developing a politically and publicly acceptable solution.

We would wish to draw your attention to the following suggestions that may well save you both money and effort in the future.

(1) Develop a 50–100 year storage facility pending research into disposal options and the emergence of a majority of public support for disposal. Until that date, all the wastes should be stored, made retrievable, and kept in continuously monitored and controlled environments.
(2) Separate storage facilities for LLW, ILW, and HLW. The greater bulk of LLW should be reduced by compaction and waste conditioning.
(3) The option not to reprocess should be reconsidered as soon as there is a real choice.
(4) The public acceptability issue should be handled by:
(a) gentle education and information campaigns;
(b) deliberate sympathy to accommodate NIMBY concerns and other non-scientific concerns;
(c) development of a 'Rolls Royce solution', at least for the first generation of facility;
(d) local compensation and other financial benefits to pay for any 'blighting' effects, maybe a surcharge on business would be appropriate;
(e) deliberate selection of low population areas to minimize the numbers of people who feel threatened. Why not use the geography of Britain to good advantage?

(5) Geographic Information System (GIS) technology should be fully utilized in order to:
(a) identify nearly optimal locations with respect to a given set of predefined and fixed siting criteria;
(b) provide a public information system capable of explaining the siting process and the nature of the constraints;
(c) justify unpopular site developments as being in the national interest.
(6) There should be some guarantee of 100 per cent safety to members of the public and a written assurance to this effect.
(7) The waste dumps should be dispersed rather than concentrated around the country with the assurance that acceptance of a radwaste site does not also bring with it a subsequent commitment to having additional nuclear developments at some future date.
(8) The minimum number of radwaste facilities necessary should be developed based on a coherent 50 year strategy for radwaste that should be revised on a regular basis.
(9) We wish that you might find time to read the attached book. We certainly enjoyed writing it and would like to think that there are some positive suggestions that may be of some assistance to you.

Good luck,
Yours sincerely,


## A Letter to the DOE

Dear Sirs,
As the principal government department, together with the Secretaries of State for Scotland and Wales, responsible for ensuring that a long term national strategy for radioactive wastes is developed and implemented, you must be very aware of the problems in this area. We wish to draw your attention to a number of questions:

Q1. Why is more emphasis not being given to long term (50 – 100 years) surface storage of LLW and ILW as a means of overcoming understandable public concerns?
Q2. Following the Chorley Report which was commissioned by your department, why are you not advocating the greater use of Geographic Information Systems (GIS) as a major site search and evaluation tool?
Q3. Why are your siting criteria so devoid of numeric values?
Q4. Why is remote siting, from a population point of view, not a feature of policy?
Q5. Why not exclude areas of high landscape value, as well as an urban nature, from the search areas?
Q6. Why not insist that the national interest argument would be better served

by requiring the developer to provide proof that the selected site(s) are among the best possible locations in Britain?

Q7. Why delegate the site choice decision to the nuclear industry? Would it not be better to select a jointly favoured location, without being constrained by matters of land ownership? Indeed, why not provide for compulsory purchase so that, in the nation's interest, the best possible facility on the best possible site can be developed?

Q8. Why not broaden the sources of independent advice?

We appreciate that radwaste siting is massively unpopular and the department obviously wishes to retain the lowest possible profile. However, this is an area which is also of massive significance and surely this requires that it commands your fullest attention.

Yours sincerely,

## A Letter to RWMAC

Dear Sirs,

Clearly you view yourselves as men of science and it is quite right and proper that you wish to base your advice to government on the best available scientific knowledge. No doubt you readily acknowledge the existence of risks in the disposal of nuclear wastes but it is also clear that you consider the risks to be both manageable and acceptable with current technology. We accept this view of the problem but would seek to condition it by the need also to take into account the broader socioeconomic milieu within which the problem is located. Our concern is that your hard 'scientific' approach has been too narrowly focused, if not positively blinkered against the importance of the non-scientific concerns and fears of the ordinary people who feel themselves threatened by radwaste dump developments. We are writing to ask why you appear to ignore these very legitimate and natural fears and why you have so far failed to see any real need to take them into account. It is of course, as you will be aware, quite feasible to design and build high quality radwaste dump facilities in good or optimal geologic locations which would also minimize the numbers of people who may feel at risk. Our work has shown that there are seemingly many potentially suitable locations that could be developed and all we ask is that the hard scientific and technical engineering siting criteria are complemented by what we would regard as commonsense remote siting principles.

There is seemingly considerable political capital to be gained by seeking remote sites as a means of introducing additional site and design independent margins of safety. Of course, there is no real scientific justification for such a recommendation; and there is no NRPB computer model that would predict the need for such an approach given the range of accident conditions that might be considered appropriate. We accept all this *but* would again emphasize the importance of gaining public acceptability by bending as far as possible to accommodate their legitimate concerns. In the real world everything does not have to have a strong scientific basis for it! There is no virtue in starting to build a radwaste dump that is liable to be cancelled prior to its completion (because of the politicalization of local public fears) or during its operational life. It has to remain acceptable not just now but for about 100 years. Narrowly based scientific approaches, that rely heavily on the faith principle, will not be sufficient to ensure that this long term political and public acceptability goal is achieved.

Like yourselves we accept the importance and urgency of finding acceptable solutions to Britain's radwaste dumping problems. We would commend to your attention, therefore, our more pragmatic approach. Why not seek to buy time by recommending an instantly acceptable long term storage solution. If not, then why not adopt a proper remote siting strategy that looks for acceptable engineering solutions with the over-riding constraint that a remote site must be used? And please, if you are going remote, then adopt a proper definition of remoteness and not merely the traditional nuclear industry view that 5 – 10 miles is sufficient. An out-of-sight, out-of-mind rule of thumb would suggest 20 – 30 miles as a workable minimum. If we can be of any assistance then please write and let us know.

Yours sincerely,

## Epilogue

So the radwaste story continues and it would be premature to predict when it might end. There will be one, maybe two, and perhaps three general elections before any national repository will be in operation and thus plenty of time and many opportunities for further political interference. In view of the past changes in policy, RWMAC(1988b) are still sufficiently pessimistic about the future to speculate on the consequences of a failure in the abilities of NIREX to secure its current objectives! Perhaps such pessimism is more than can at present be justified; perhaps RWMAC are still smarting at the 1987 abandonment of the four near surface sites without themselves being consulted; perhaps they believe they have more 'power' than they in reality possess. Alternatively, perhaps they are anticipating another 'change in policy'. The radwaste dump problem is of course largely artificial. It was created by the government's determination to dispose rather than store the wastes. Change this policy requirement, and long term but not indefinite storage becomes extremely attractive. Perhaps the government was never really bothered about any disposal option and this policy requirement might have been manufactured by RWMAC themselves, merely being passed on to others as government policy. Certainly, RWMAC have followed a very strong and very pro-nuclear industry line; it is almost as if they are dominated by or acting as a voice of the old-time 'hardliners'. It is perhaps a pity that NIREX were not given a freer hand to investigate and solve the radwaste dump problem themselves without having to be so seemingly dependent on these 'policy' factors. We wish them luck and no doubt many future generations of Britains will hope even more fervently than we do that they succeed.

# References

*ATOM* (1983). 'British waste sites announced'. *ATOM* **326**, December 1983.

*ATOM* (1987). 'UK opts for deep disposal of low level waste'. *ATOM* **368** 24.

Baker, J. (1987). Letter to secretary of state for environment. May 1987.

Bath, A.H., George, I.A., Milodowski, A.E. and Darling, W.G. (1985). *Long Term Effects on Potential Repository Sites: occurrence and diagensis of anhydrite.* BGS Fluid Processes Research Group Report No. FLPU 85–12, October 1985, BGS/NERC.

Beale, H. (1985). 'The meticulous five–step search for a site.' *PlainTalk* May 1985, 4–5.

Beale, H. (1987) The assessment of potentially suitable repository sites. in *The Management and Disposal of Intermediate and Low Level Radioactive Waste.* London, Mechanical Engineering Publications Ltd. 11–18.

Black, D. (1984) *Investigation of the possible increased incidence of cancer in West Cumbria.* Report of the Independent advisory group, London, HMSO.

Bloomfield, A.M. (1988). 'Progress in the development of fast reactors' *ATOM* **382** 2–6.

Blowers, A., Lowry, D. (1985). 'Out of Sight, Out of Mind.' (mimeo) Paper presented at the IBG Conference, Leeds, January 1985.

Blowers, A. and Pepper, D. (1987). *Nuclear Power in Crisis*, London, Croom Helm.

Bond (1987). Britain Opposed to Nuclear Dumping (mimeo).

BNFL (1985). Memorandum to the House of Commons Environment Committee First Report on the Disposal of Radioactive Wastes. London, HMSO.

BNFL (1985). *Nuclear Fuel Reprocessing Technology*, Risley, BNFL.

BNFL (1986). 'Drigg' information bulletin.

Buchan-Smith, A. (1983) *Hansard* (37), London, HMSO, 144–45.

Carver, S. (1989). 'Where to bury our nuclear waste?' Paper presented at IBG annual conference, Coventry, January 1989.

CEGB (1985). Hinckley Point C public inquiry. Statement of case.

Chapman, N.A., McEwan, T.J. and Beale, H., (1986). 'Geological Environments for Deep Disposal of Intermediate Level Wastes in the UK.' Paper presented at IAEA International Symposium on the Siting, Design and Construction of Underground Repositories for Radioactive Wastes, 3–7 March 1986, Hanover, IAEA Report No. IAEA–SM–289/37.

Charlesworth, F.R. (1985). 'Control of radioactive wastes from reprocessing irradiated fuel; some reflections on recent experience,' *Radioactive Waste Management: technical hazards and public acceptance.* Conference Report. London, Oyez Scientific and Technical Services Ltd.

Charlesworth, F.R. and Gronow, W.S. (1967). A summary of experience in the practical application of siting policy in the UK. in *Containment and Siting of Nuclear Power Plants.* IAEA, Vienna, 143–170.

Cook, F. (1983) *Hansard* (47) 25 October, London, HMSO, 157–58.

Cook, F. (1983) *Hansard* (49) 30 November, London, HMSO, 501–504.

Cook, F. (1984) *Hansard* (59) 3 May, London, HMSO, 144–45.

Coopers & Lybrand Associates (1986). *Killingholme: the development potential.* Report commissioned by Humberside County Council, October 1986.

Davies, A.W. (1987). Deep Repository Concepts for Long-Lived Radioactive Wastes, in *Management and disposal of intermediate and low level radioactive waste.* London, Mechanical Engineering Publications,.33–38.

DOE (1977). *Nuclear Power and the Environment,* Cmnd 6820, London, HMSO.

DOE (1982). *Radioactive Waste Management,* Cmnd 8607, London, HMSO.

DOE (1983). *Disposal facilities on land for low and intermediate level radioactive waste: draft principles for the protection of the human environment.* London, HMSO.

DOE (1984). *Statement of principles to be taken into account when considering authorising land disposal facilities for low and intermediate level radio-active waste* DOE/5-13, Sizewell B Power Station Public Inquiry. London, HMSO.

DOE (1984). *Radioactive Waste Management: the national strategy.* London, HMSO.

DOE (1985). *Disposal facilities on land for low and intermediate level radioactive wastes: principles for the protection of the human environment.* London, HMSO.

DOE (1985–6). *Minutes of evidence by Department of Environment Officials, Vol. 2* Environment Committee Report, London, HMSO.

DOE (1986a). *Radioactive Waste: The Government's Response to the Environment Committee's Report.* Cmnd 9852, London, HMSO.

DOE (1986b). *Assessment of Best Practicable Environmental Options (BPEOs) for management of low and intermediate level solid radioactive wastes.* London, HMSO.

DOE (1987). *Handling Geographic Information:* Report of the Committee of Enquiry chaired by Lord Chorley. London, HMSO.

DOE (1988). *The 1987 United Kingdom Radioactive Waste Inventory.* Report prepared for the UK NIREX Ltd, and the DOE. Electrowatt Engineering Services (UK) Ltd.

Environmental Resources Ltd., (1987). *The Disposal of Radioactive Waste in Sweden, West Germany and France.* ERL Report prepared for Bedfordshire, Humberside and Lincolnshire County Councils. January, London, ERL.

ERAU (1988). *Responses to the Way Forward,* Environmental Risk Assesment Unit, School of Environmental Sciences, University of East Anglia.

ESRI (1987). *Arc/Info: the Geographic Information Systems software, Users Guide Volume 1,* California, Environmental Systems Research Institute.

Expert Group of Government Departments, (1979). *The Control of Radioactive Wastes: a review of Cmnd 884,* London, HMSO.

Fernie, J. and Openshaw, S. (1986). 'Who wants nuclear waste?' *The Geographical Magazine.* LVIII, 63–69.

Fyfe, W.F., Babuska, V., Price, N.J., Schmid, E., Tsang, C.F., Uyeda, S., and Velde, B. (1984). The geology of nuclear waste disposal. *Nature* 310, August London.

Ginniff, M.E., (1985). 'The implementation of UK policy and strategy on radioactive waste disposal.' *Nuclear Energy* 24(2) pp.99–104.

Ginniff, M.E., (1987). 'The characteristics of disposal sites and repositories.' *Nuclear Energy* 24(2) 5–10.

Ginniff, M.E. and Phillipson, D.L., (1984). 'The disposal of solid radioactive waste to land sites in the UK.' *The Nuclear Engineer* 25(5) 193–197.

Griffin, J.R., Hackney, S., Heafield, W. and Richardson, J.A. (1982). *An engineering design study for storage and disposal of intermediate level waste.* UKAEA Report ND–R777.

Grove, J.R., and Hickford, G.E. (1984). 'NIREX plans for the transport and disposal of decommissioning wastes.' Paper presented at Decommissioning of Radioactive Facilities Seminar organised by the Institute of Mechanical Engineers, 7 November, 29–34.

*Hansard* (1986). Nuclear Waste: statement by DOE on Drigg disposal limits at Drigg in answer to question by Douglas Hogg, 27 June 1986.

Holliday, F.G.T (1984). *Report of the Independent Review of Disposal of Radioactive Waste in the North East Atlantic.* London, HMSO.

House of Commons Environment Committee (1986). *First Report from the Environment Committee on Radioactive Waste.* London, HMSO.

Humberside County Council (1985). *Humberside Structure Plan: radioactive waste disposal facilities; explanatory memorandum, and consultation and participation.*

IAEA (1983). *Disposal of low and intermediate level solid radioactive wastes in rock cavities.* Safety Series No. 59. Vienna, IAEA.

ICI (1983). ICI's position on the possible use of Billingham anhydrite mine for radioactive waste. Public statement issued by ICI, 25 October 1983.

ICI (1984). ICI's position on the possible use of Billingham anhydrite mine for Radioactive waste. Public statement issued by ICI, 5 March 1984.

Jones, S.R. (1986). Statement to the Environmental Health Sub-committee of the Sellafield local Liaison Committee on the geology of Drigg. BNFL, 17 April 1986.

Kemp, R., O'Riordan, T. and Purdue, M. (1986). 'Environmental politics in the 1980s: the public examination of radioactive waste disposal.' *Policy and Politics*. 4(1) 9–25.

Layfield, F. (1987). *Sizewell B Public Inquiry*. Department of Energy, London, HMSO.

Lewis, J.B. (1983). 'The case for deep-sea disposal of low-level solid radioactive wastes.' *Nuclear Energy* 22(1) 47–51.

Macgill, S. (1987). *The Politics of Anxiety*. London, Pion.

McEwen, T.J. (1985). *Preliminary Assessment of the Geology for the Proposed Site Investigation at Bradwell, Essex*. Report prepared for UK NIREX Ltd., September 1985, NIREX report No. 21.

McEwen, T.J. (1986a). 'Comments on Robins' *Report on Bradwell in Relation to the Disposal of Low Level Radioactive Wastes on the Site*. Statement prepared for NIREX, BGS, January.

McEwen, T.J. (1986b). *Preliminary Assessment of the Geology for the Proposed Site investigation at Fulbeck, Lincolnshire*. Report prepared for UK NIREX Ltd., May, NIREX Report No. 18.

McInerney, P.T. (1987). 'The overall strategy for radioactive waste management.' *Nuclear Energy* 24(2) 1–4.

Milodowski, A.E., Bloodworth, A.J. and Wilmot, R.D. (1985). Long term effects on potential repository sites: The alterations of the Lower Oxford Clay during weathering. British Geographical Society. Report No. FUPU 85–13 September. NERC/BGS.

Morris, C.H. (1979). *Report on Abandoned mineral Workings and Possible Surface Instability Problems*. County of Cleveland, Department of the County Survey and Engineer, December.

NEI (1985). 'Billingham dumped as search for waste site widens.' *Nuclear Engineering International* 30 (366), 12.

Nuclear Regulatory Commission (1987). *Guidance for selecting sites for the disposal of low level waste*. Washington DC, NRC Regulatory Guide.

NIREX (1983). *The Disposal of Low and Intermediate Level Radioactive Wastes: the Elstow storage depot, a preliminary project statement*. UK NIREX Ltd., October.

NIREX (1984). *UK NIREX Ltd Site Identification Procedures: information for local authorities*. Harwell, NIREX.

NIREX (1985). *Third Report*. Harwell, NIREX.

NIREX (1986a). NIREX Site Announcements: information from NIREX. UK NIREX Ltd., 25 February.

NIREX (1986b). *Disposal of Low Level Radioactive Waste: the Bradwell site, a preliminary project statement*. UK NIREX Ltd., October.

NIREX (1986c). *The Disposal of Low Level Radioactive Waste: the Elstow storage depot, a preliminary project statement*. UK NIREX Ltd.

NIREX (1986d). *The Disposal of Low Level Radioactive Waste: Fulbeck, a preliminary project statement*. UK NIREX Ltd.

NIREX (1986e). *The Disposal of Low Level Radioactive Waste: S. Killingholme, a preliminary project statement.* UK NIREX Ltd.

NIREX (1986f). *Preliminary Safety Report and Environmental and Radiological assessment: the sites under investigation for the disposal of low level waste.* NIREX Ltd., November.

NIREX (1987). Press Release, 1 May.

NIREX (1987). *The Way Forward: a discussion document.* UK NIREX Ltd., November.

OECD/NEA (1985). *North East Atlantic five year site suitability review.* OECD.

Openshaw, S. (1986). *Nuclear Power: siting and safety.* London, Routledge.

Openshaw, S. (1988). 'Making nuclear power more publicly acceptable.' *Journal of British Nuclear Energy Society* 27, 131–136.

Openshaw, S. and Fernie J. (1986). Computer techniques for identifying and evaluating potential radioactive waste disposal sites. in *Radioactive Waste.* 1st report of the Environment Committee, 3, 645–647.

Openshaw, S., Wymer, C. and Charlton, M. (1986). A geographical information and mapping system for the BBC Domesday optical discs. *Transactions of the Institute of British Geographers.* New Series 11, 315–325.

Openshaw, S., Charlton, M., Lugnor, C., Galt, C. (1987). 'A Mark I geographical analysis machine for the automated analysis of data sets.' *International Journal of Geographical Information Systems* 1, 335–358

Openshaw,S., Carver, S.J. and Fernie, J. (1988). *Siting issues in the disposal of low and intermediate level radioactive wastes: the role of geographic information systems.* Northern Regional Research Laboratory internal report No. 17. Newcastle University.

Openshaw, S., Charlton, M., Galt, A.W., Birth, J.M. (1988). 'An investigation of Leukaemia clusters by use of a geographical analysis machine.' *The Lancet*, 6 February 272–273

O'Riordan, T. (1986). *Memorandum to the House of Commons Environment Committee first report on the disposal of radioactive wastes.* London, HMSO.

O'Riordan, T. (1987). Prospects for the nuclear debate in the UK. in Blowers, A., and Pepper, D. (1987). *Nuclear Power in Crisis.* London, Croom Helm.

*PlainTalk* (1983 to 1988). UK NIREX Ltd information paper.

Quick, J. (1988). 'The problem of radioactive waste', *ATOM* 379, 19–21.

Radioactive Waste Management Advisory Committee (RWMAC) Annual Reports, 1980–1988.

RWMAC (1988a). *Report of the RWMAC subgroup on the NIREX proposals for deep site investigations.* London, HMSO.

RWMAC (1988b). *Ninth Annual Report.* London, HMSO.

Rhind, D. and Mounsey H. (1986). The land and people of Britain: a Domesday record. *Transactions of the Institute of British Geographers.* New Series 11, pp.296–304.

Roberts, L.E.J. (1979). Radioactive waste disposal – policy and perspectives. in *Nuclear Energy* 18, pp.85–100.

Robins, N.S. (1980). *The Geology of Some United Kingdom Nuclear Sites Related to the Disposal of Low and Intermediate Level Radioactive Wastes.* IGS, NERC Report No. ENPU 80–9, Harwell, Environmental Protection Unit, IGS, April.

Royal Commission on Environmental Pollution (1976). *Sixth Report: Nuclear Power and the Environment.* (The Flowers Report), London, HMSO.

Royal Commission on Environmental Pollution (1984) *Tenth Report: Tackling Pollution – experience and prospects,* London, HMSO.

Saunders, P. (1987). *An Outline of NIREX's Research and Safety Assessment Programmes. NIREX Radioactive Waste Disposal: safety studies.* Report No. NSS/G101, November.

Schneider, K.R. (1985). 'The Treatment and Preparation for Disposal of High Level and intermediate level wastes.' Radioactive Waste Management: Technical Hazzards and Public Acceptance. Conference. 5–6 March 1985. London, Oyez Scientific & Technical Services Ltd. 155–164.

Williams, G.M. (1985). *Preliminary Assessment of the Geology and the Hydrogeology for the proposed site investigation at the Elstow storage depot, Bedfordshire.* Report prepared for the UK NIREX Ltd., January, NIREX Report No. 22.

Williams, G.M., Stuart, A. and Holmes, D.C. (1985). *Investigations of the geology of the low level radioactive waste burial site at Drigg, Cumbria.* BGS Report 17(3), London, HMSO.

# Index

ALARA, 60, 61
ALARP, 35, 60, 61
Altnabreac, 35, 113, 114, 144, 153, 160, 165, 167, 171, 189, 191
Amersham International, 34

Bedfordshire/Billingham against Nuclear Dumping (BAND), 54, 56, 75, 82
Berkeley, 34
Best Practicable Environmental Option (BPEO), 48, 52, 53-57, 67, 81, 83, 85, 88, 98, 99, 115, 181, 192, 200
Billingham, 35, 37, 53, 55, 56, 57, 66, 67, 69, 73, 74, 75, 76, 77, 79, 105, 106, 113, 114, 120, 121, 131, 137, 153, 158, 160, 167, 171, 189, 191, 202
Bradwell, 34, 35, 37, 56, 80, 82, 86, 92, 108, 109, 112, 113, 153, 160, 167, 171, 175, 191, 202
Britain Opposed to Nuclear Dumping (BOND), 81, 82
British Nuclear Fuels Limited (BNFL), IX, 11, 22, 24, 26, 31, 35-37, 40, 47, 63, 66, 68, 70, 71, 99-104, 199
British Geological Society (BGS), 102, 103, 107, 138, 141, 204

Calder Hall, 21
Capenhurst, 34
Central Electricity Generating Board (CEGB), 15, 22, 26-28, 33, 40, 50, 53, 68, 71-73, 83, 84, 107, 109-112, 116, 117, 119, 120, 125, 175
Chapelcross, 34
Chernobyl, 7, 8, 13, 16, 21, 29, 104, 121, 179
Chorley Report, 145, 146, 195, 200

County Councils Coalition (CCC), 82-85, 114
Culham, 34

Decomissioning, 2-5, 20, 21, 33, 34, 37, 58, 69, 99, 111, 119, 154, 191
Deep Cavity Disposal (DCD), 93, 94, 137, 142, 143
Department of the Environment (DOE), 13, 22, 29, 34, 37, 47, 49, 51-57, 59, 62, 64, 66, 67, 73, 81, 87, 95, 98, 101, 111, 115, 117, 120-125, 135, 145, 146, 173, 174, 177, 180-183, 187, 192, 195, 196, 200
Dounreay, 24, 33-35, 60, 63, 70, 90, 113, 114, 143, 144, 158, 160, 165, 167, 170, 171, 189, 191
Drigg, 2, 21, 32, 33, 45, 47, 60, 69-71, 80, 84, 92, 93, 99-104, 107, 113, 127, 158, 160, 165, 167, 171, 189, 191, 201, 204
Dungeness, 34

Elstow, 35, 37, 53, 56, 57, 66, 67, 73, 76, 78-82, 86, 92, 106-108, 112, 113, 115, 116, 121, 150, 153, 160, 167, 170, 171, 175, 191, 204
Engineered Trench Disposal (ETD), 81, 92, 93, 94, 107, 170
Essex Against Nuclear Dumping (EAND), 82
Euratom, 46

Fulbeck, 35, 37, 56, 80, 82, 86, 92, 109, 112, 113, 153, 160, 167, 171, 175, 191

205